RESEARCH AND PERSPECTIVES IN ALZHEIMER'S DISEASE

Fondation Ipsen

J. Cummings J. Hardy M. Poncet
Y. Christen (Eds.)

Genotype – Proteotype – Phenotype Relationships in Neurodegenerative Diseases

With 20 Figures and 21 Tables

Springer

Jeffrey L. Cummings, M.D.
Reed Neurological Research Center, UCLA
710 Westwood Plaza
Box 951769
Los Angeles, CA 90095-1769
USA
E-mail: jcummings@mednet.ucla.edu

John Hardy, Ph.D.
Laboratory of Neurogenetics
National Institute on Aging
National Institutes of Health
Building 35, Rooms 1A1015
MSC 3707, Bethesda, MD 20892
USA
E-mail: hardyj@mail.nih.gov

Michel Poncet, M.D.
Neurologie
CHU La Timone
Bd. Jean Moulin
13385 Marseille Cedex 05
France
E-mail: miponcet@mail.ap-hm.fr

Yves Christen, Ph.D.
Fondation IPSEN
Pour la Recherche Thérapeutique
24, rue Erlanger
75781 Paris Cedex 16
France
e-mail: yves.christen@beaufour-ipsen.com

ISSN 0945-6066
ISBN 3-540-24835-8 Springer-Verlag Berlin Heidelberg New York

Library of Congress Control Number: 2005920457

Springer-Verlag is a part of Springer Science+Business Media

springeronline.com

Printed in Germany

Editor: Dr. Rolf Lange, Heidelberg
Desk Editor: Anne Clauss, Heidelberg
Production: PRO EDIT GmbH, 69126 Heidelberg, Germany
Cover design: design & production, 69121 Heidelberg, Germany
Typesetting: SDS, Leimen, Germany
Printed on acid-free paper 27/3150Re – 5 4 3 2 1 0

Preface

Neurodegenerative diseases, including Alzheimer's disease and Parkinson's disease, are common afflictions of the elderly and account for a major amount of morbidity and mortality in older individuals. Less common conditions such as frontotemporal dementias, dementia with Lewy bodies, progressive supranuclear palsy, corticobasal degeneration, and Huntington's disease also are neurodegenerative diseases that cause progressive neurological dysfunction and eventual death. Neurodegenerative diseases produce cognitive, functional and behavioral disturbances, place tremendous demands on family caregivers, may lead to institutionalization, and are fatal. Many neurodegenerative diseases are age-related and the increasing size of the aged population will inevitably result in a greater number of patients suffering from neurodegenerative conditions unless means of preventing, delaying or treating these disorders are found. There is an urgent need to identify tractable therapeutic targets within the pathophysiology of neurodegeneration.

Abnormalities of protein metabolism, including protein misfolding, are increasingly recognized as central to the mechanisms of most neurodegenerative processes. Amyloid beta protein production and accumulation is considered central to the pathophysiology of Alzheimer's disease; disturbances of proteasome and alpha- synuclein metabolism are present in Parkinson's disease, dementia with Lewy bodies, and multi-system atrophies; disorders of tau or ubiquitin metabolism have been identified in frontotemporal dementias, progressive supranuclear palsy, and corticobasal degeneration. Abnormalities of huntingtin are characteristic of Huntington's disease. Other neurodegenerations have uniquely associated protein abnormalities. Prion disorders also result in abnormal protein deposits and self-propagating protein misfolding appears to be the major pathophysiologic event in prion diseases. Recent advances in understanding the role of protein dysmetabolism in neurodegeneration was the theme of the Fondation IPSEN meeting (September 13, 2004) addressing Genotype-Proteotype-Phenotype relationships. Experts from international laboratories contributed to the conference and the current volume to produce a comprehensive overview of the role of protein misfolding in neurodegeneration. Links between genotype and protein characteristics and between proteotype and clinical phenomenology were discussed across diseases categories. Progress in understanding the role of abnormalities of protein metabolism may lead to the identification of biological markers relevant to disease monitoring and to the development of new therapeutic agents capable of modifying and ameliorating basic neurodegenerative mechanisms.

Acknowledgement
The editors wish to thank Jacqueline Mervaillie for organizing the meeting in Paris and Astrid de Gerard, Katrina Kerr and Mary-Lynn Gage for the editing of the book.

Los Angeles, California,
January 2005

Jeffrey L. Cummings

Contents

Neurodegenerative Disorders as Proteinopathies:
Phenotypic Relationships
Jeffrey L. Cummings 1

Towards a Molecular Classification of Neurodegenerative Disease
John Hardy 11

Racial and Ethnic Influences on the Expression
of the Genotype in Neurodegenerative Diseases
Katrina Gwinn-Hardy 25

Causes and Consequences of Oxidative Stress
in Neurodegenerative Diseases
Domenico Praticò 37

Early Onset Familial Alzheimer's Disease:
Is a Mutation Predictive of Pathology?
Douglas Galasko 45

Identification of Genes that Modify the Age of Onset
in a Large Familial Alzheimer's Disease Kindred
*Alison M. Goate, Pau Pastor, Catherine M. Roe, Joanne Norton,
Sumi Chakraverty, Francisco Lopera, Kenneth S. Kosik* 61

Variable Phenotype of Alzheimer's Disease with Spastic Paraparesis
*Helena Karlstrom, William S. Brooks, John B.J. Kwok,
Jillian J. Kril, Glenda M. Halliday, Peter R. Schofield* 73

Presenilin Mutations: Variations in the Behavioral Phenotype
with an Emphasis on the Frontotemporal Dementia Phenotype
Neill R. Graff-Radford 93

Frontotemporal Dementias: Genotypes and Phenotypes
Virginia M.-Y. Lee 103

Chromosome 17-Linked Frontotemporal Dementia
with Ubiquitin-Positive, tau-Negative Inclusions
*Rosa Rademakers, Julie van der Zee, Samir Kumar-Singh, Bart Dermaut,
Marc Cruts, Christine Van Broeckhoven* 117

Variations of the Phenotype in Frontotemporal Dementias
N. L. Foster, R. E. Koeppe, B. J. Giordani, T. E. Nichols, A. P. Lieberman 139

Phenotype/Genotype Correlations in Parkinson's Disease
*Alexis Brice, Ebba Lohmann, Pablo Ibanez, Magali Periquet, Sophie Laine,
Béatrice Debarges, Suzanne Lesage, and Alexandra Dürr* 153

Subject Index 165

Contributors

Brice, Alexis
INSERM Unit 289 and Département de Génétique,
Cytogénétique et Embryologie, Hôpital de la Pitié-Salpêtrière, AP-HP
47 boulevard de l'Hôpital, 75651 Paris Cedex 13
France

Brooks, William S
Prince of Wales Medical Research Institute, University of New South Wales,
Barker Street, Randwick, Sydney
Australia

Chakraverty, Sumi
Department of Psychiatry, Washington University School of Medicine,
Campus Box 8134, 660 South Euclid Avenue, St. Louis, MO 63110
USA

Cruts, Marc
Department of Molecular Genetics,
Flanders Interuniversity Institute for Biotechnology (VIB),
University of Antwerp (UA), 2610 Antwerpen
Belgium

Cummings, Jeffrey L.
Departments of Neurology and Psychiatry and Biobehavioral Sciences,
David Geffen School of Medicine at UCLA, Los Angeles, CA, USA,
Reed Neurological Research Center, UCLA,
710 Westwood Plaza, Box 951769, Los Angeles, CA 90095-1769
USA

Debarges, Béatrice
INSERM Unit 289, Département de Génétique, Cytogénétique et Embryologie
Hôpital de la Pitié-Salpêtrière, AP-HP, 47 boulevard de l'Hôpital,
75651 Paris Cedex 13
France

Dermaut, Bart
Department of Molecular Genetics,
Flanders Interuniversity Institute for Biotechnology (VIB),
University of Antwerp (UA), 2610 Antwerpen,
Belgium

Dürr, Alexandra
INSERM Unit 289, Département de Génétique
Cytogénétique et Embryologie, Hôpital de la Pitié-Salpêtrière, AP-HP,
47 boulevard de l'Hôpital, 75651 Paris Cedex 13
France

Foster, Norman L.
Department of Neurology, University of Michigan,
1920 Taubman Center
1500 E. Medical Center Dr. Ann Arbor, MI 48109-0316,
USA

Galasko, Douglas
Professor, Department of Neurosciences, UCSD,
and Staff Physician, Veterans Affairs Medical Center,
3350 La Jolla Village Drive, San Diego CA 92161,
USA

Giordani, B. J.
Department of Psychiatry, University of Michigan
Ann Arbor, MI, 48109,
USA

Goate, Alison M.
Department of Psychiatry, Washington University School of Medicine,
Campus Box 8134, 660 South Euclid Avenue, St. Louis, MO 63110,
USA

Graff-Radford, Neill R.
Mayo College of Medicine, 4500 San Pablo Road, Jacksonville, Florida 32224
USA

Gwinn-Hardy, Katrina
National Institute of Neurological Disorders and Stroke,
National Institutes of Health, Bethesda, MD 20892
USA

Halliday, Glenda M
Prince of Wales Medical Research Institute, University of New South Wales,
Barker Street, Randwick , Sydney,
Australia

Hardy, John
Laboratory of Neurogenetics , National Institute on Aging
National Institutes of Health, Building 35, Room 1A1015, MSC 3707
Bethesda, MD 20892
USA

Ibanez, Pablo
INSERM Unit 289, Département de Génétique, Cytogénétique et Embryologie,
Hôpital de la Pitié-Salpêtrière, AP-HP,
47 boulevard de l'Hôpital, 75651 Paris Cedex 13
France

Karlstrom, Helena
Garvan Institute of Medical Research, Sydney
Australia

Koeppe R. E.
Department of Radiology, University of Michigan, Ann Arbor, MI 48109
USA

Kosik, Kenneth S.
Center for Neurological Diseases, Harvard Medical School,
Brigham and Women's Hospita, Boston, MA
USA

Kril, Jillian J
Centre for Education and Research on Ageing, University of Sydney, Sydney
Australia

Kumar-Singh, Samir
Department of Molecular Genetics,
Flanders Interuniversity Institute for Biotechnology (VIB),
University of Antwerp (UA), 2610 Antwerpen
Belgium

Kwok, John B.J.
Garvan Institute of Medical Research, Sydney
Australia

Laine, Sophie
INSERM Unit 289, Département de Génétique, Cytogénétique et Embryologie,
Hôpital de la Pitié-Salpêtrière, AP-HP,
47 boulevard de l'Hôpital, 75651 Paris Cedex 13
France

Lee, Virginia M.-Y.
Center for Neurodegenerative Disease Research,
Department of Pathology and Laboratory Medicine and the Institute on Aging,
University of Pennsylvania School of Medicine,
Philadelphia, PA, 19104-4283
USA

Lesage, Suzanne
INSERM Unit 289, Département de Génétique, Cytogénétique et Embryologie,
Hôpital de la Pitié-Salpêtrière, AP-HP
47 boulevard de l'Hôpital, 75651 Paris Cedex 13
France

Lieberman, A. P.
Department of Pathology, University of Michigan, Ann Arbor, MI 48109
USA

Lohmann, Ebba
INSERM Unit 289, Département de Génétique, Cytogénétique et Embryologie,
Hôpital de la Pitié-Salpêtrière, AP-HP,
47 boulevard de l'Hôpital, 75651 Paris Cedex 13
France

Lopera, Francisco
Grupo de Neurociencias, Facultad de Medicina,
Universidad de Antioquía, Medellín
Colombia

Nichols, T. E.
Department of Biostatistics, University of Michigan, Ann Arbor, MI, 48109,
USA

Norton, Joanne
Department of Psychiatry, Washington University School of Medicine,
Campus Box 8134, 660 South Euclid Avenue, St. Louis, MO 63110
USA

Pastor, Pau
Department of Psychiatry, Washington University School of Medicine,
Campus Box 8134, 660 South Euclid Avenue, St. Louis, MO 63110
USA

Periquet, Magali
INSERM Unit 289, Département de Génétique, Cytogénétique et Embryologie,
Hôpital de la Pitié-Salpêtrière, AP-HP,
47 boulevard de l'Hôpital, 75651 Paris Cedex 13
France

Praticò, Domenico
Center for Experimental Therapeutics, Department of Pharmacology,
University of Pennsylvania, School of Medicine,
John Morgan Building, room 124, Philadelphia, PA, 19104
USA

Rademakers, Rosa
Department of Molecular Genetics,
Flanders Interuniversity Institute for Biotechnology (VIB),
University of Antwerp (UA), 2610 Antwerpen
Belgium

Roe, Catherine M.
Division of Biostatistics and Alzheimer's Disease Research Center,
Washington University School of Medicine, St. Louis, MO
USA

Schofield, Peter R.
University of New South Wales, Sydney Australia,
Garvan Institute of Medical Research, Sydney, Australia,
Prince of Wales Medical Research Institute,
Barker Street, Randwick, Sydney 2031
Australia

Van Broeckhoven, Christine
Department of Molecular Genetics (VIB8), University of Antwerp (UA),
Universiteitsplein 1, B-2610 Antwerpen
Belgium

van der Zee, Julie
Department of Molecular Genetics,
Flanders Interuniversity Institute for Biotechnology (VIB),
University of Antwerp (UA), 2610 Antwerpen
Belgium

Neurodegenerative Disorders as Proteinopathies: Phenotypic Relationships

Jeffrey L. Cummings[1]

Summary

The phenotype of neurodegenerative diseases is defined by the distribution of cellular changes, which in turn depends on vulnerability to specific abnormalities of protein metabolism. Distinctive features have been identified that assist in the recognition of specific diseases or specific types of abnormalities of protein metabolism that are shared by several neurodegenerative disorders. Identification of the earliest manifestations of the phenotype will facilitate early implementation of therapy and improve understanding of protein metabolism abnormalities and their secondary consequences. This will assist in drug development and effective therapeutics for neurodegenerative disorders.

Introduction

Abnormalities in the production, aggregation, metabolism, and removal of proteins are increasingly recognized as a shared feature of neurodegenerative disorders (Cummings 2003a; Hardy and Gwinn-Hardy 1998; Lovestone and McLoughlin 2002; Soto 2003; Taylor et al. 2002; Trojanowski and Lee 2000; Ross and Poirier 2004). Identification of protein misfolding across a wide variety of neurodegenerative diseases has stimulated new areas of research and may lead to promising therapeutic interventions.

Each current distinguishable neurodegenerative disease is characterized by the accumulation of a unique protein type (Table 1). In addition, each disease is characterized by a uniquely vulnerable cell population. Protein disturbance occurs in the most vulnerable (or least resistant) cell type, producing local dysfunction and leading to an observable clinical phenotype. Thus, the well characterized phenotype can be viewed as relevant to the at-cell-risk population and to the proteotype of neurodegenerative disease. This unique linkage between proteotype, vulnerable cell population, and phenotype is the topic of this review.

Abnormalities of three proteins account for more than 90% of all neurodegenerative disorders. These include: amyloid beta protein, present in Alzheimer's disease (AD), Down Syndrome, and the common form of dementia with Lewy

[1] Departments of Neurology and Psychiatry and Biobehavioral Sciences, David Geffen School of Medicine at UCLA, Los Angeles, CA, USA.

Cummings et al.
Genotype – Protéotype – Phenotype
Relationships in Neurodegenerative Diseases

Table 1. Protein disturbances associate with neurodegenerative disease.

Disease	Protein Abnormalities
Alzheimer's disease	Amyloid-β protein Tau protein
Parkinson's disease	Alpha-synuclein
Dementia with Lewy bodies (pure form)	Alpha-synuclein
Dementia with Lewy bodies (common form)	Alpha-synuclein Amyloid-β protein
Multiple system atrophy (indicate Shy-Drager disorder, olivopontocerebellar degeneration, striatonigral degeneration)	Alpha-synuclein
Frontotemporal dementia	Tau protein
Progressive supranuclear palsy	Tau protein
Corticobasal degeneration	Tau protein
Guamanian ALS *-parkinsonism dementia complex	Tau protein
Pallidonigral degeneration	Tau protein
Creutzfeldt-Jakob disease	Prion protein
Gerstmann-Straussler-Scheinker disease	Prion protein
Fatal familial insomnia	Prion protein
Kurn	Prion protein
Familial ALS*	Superoxide dismutase
Huntington's disease	Huntingtin
Spinocerebellar ataxia	Ataxin
Dentato-rubro-pallidoluysian atrophy	Atrophin
Familial encephalopathy with neuroserpin inclusion bodies	Neuroserpin
Adult polyglucosan body disorder	Polyglucosan Alpha-synuclein
Lafora progressive myoclonus epilepsy	Laforin

*ALS-amyotrophic lateral sclerosis

bodies; tau protein disturbances, found in frontotemporal dementia, progressive supranuclear palsy, corticobasal degeneration, Guamanian amyotrophic lateral sclerosis (ALS)-parkinsonism-dementia complex, pallidopontonigral generation, and AD; and alpha-synuclein disorders, present in dementia with Lewy bodies (pure and common forms), Parkinson's disease, the multiple system atrophies (including Shy-Drager disorder, olivopontocerebellar atrophy, striatonigral degeneration) and neurodegeneration with brain iron accumulation (Hallervorden-

Table 2. Differences between inherited and sporadic forms of neurodegenerative disorders.

Inherited	Sporadic
Mutation present	Mutation absent
Rare	Common
Early onset	Late onset
More rapid progression	Slower progression
Protein production abnormalities	Protein accumulation abnormalities

Spatz syndrome; Cummings 2003b). Inherited and Sporadic Forms of Neurodegenerative Diseases.

Many neurodegenerative diseases exist in two forms: an inherited form, in which the genetic transmission of the disease can be delineated, and idiopathic forms of the disease, in which genetic influences are subtle or absent. The pathology of the two forms is typically identical or very similar. Investigation of the molecular genetics of the inherited forms has provided a window into the molecular biology of both the inherited and the sporadic forms of neurodegenerative diseases. In addition, characterization of mutations associated with inherited forms of neurodegenerative disorders has provided an opportunity to create transgenic animals that mimic some aspect of these diseases.

Clinical features may differ between the inherited and sporadic forms of neurodegenerative diseases (Table 2). Inherited forms of neurodegenerative diseases are much rarer than the sporadic forms of the disorder; for example, autosomal dominant AD accounts for approximately 3% of all cases of AD (Blacker and Tanzi 2000) whereas 97% are of the sporadic type. In addition, inherited forms of neurodegenerative disorders tend to begin much earlier in life than the sporadic forms (Bird 1999). Autosomal dominant AD often begins between ages 50 and 60; the sporadic form typically begins after age 70. Disease duration also tends to be shorter in inherited forms of neurodegenerative disorders than in sporadic forms. Early onset and shorter duration typically are correlated (Bird 1999). Within families, distinctive phenotypes may be recognized such as familial cases of AD with more psychosis or more language disturbance, but these features have not been shown to be characteristic of early onset AD as a class.

Phenotype of Frontotemporal Dementia and Other Tauopathies

Tauopathies show a remarkable predilection for involvement of frontal lobes and related structures of frontal-subcortical circuits. Such a grouping implies that structures that participate in common functions share a common molecular vulnerability to specific proteotype abnormalities. The frontotemporal lobar degenerations (including frontotemporal dementia, semantic dementia, and primary progressive aphasia) preferentially involve the frontal lobes and anterior temporal lobes; corticobasal degeneration involves the basal ganglia and diverse regions of the cerebral cortex, including the frontal lobes; progressive supranuclear palsy involves the globus pallidus and thalamus (Hauw et al. 1994; Dickson et al. 2000; Munoz 1998).

There are distinctive phenotypic manifestations of dysfunction of frontal-subcortical circuits (Cummings 1993). Dorsolateral prefrontal subcortical circuits mediate executive function involving planning, programming, implementing, and adjusting traditional activity. Abnormalities in this circuitry are typically manifested by disturbances in abstraction, judgment, insight, empathy, verbal fluency, motor programming, the ability to ignore the routine in favor of novel responses, attention, and concentration. Perseveration, poor recall with intact recognition, reduced verbal and nonverbal fluency, and abnormalities of initiation, utilization, and imitation behavior are typical of patients with disturbances of the dorsolateral prefrontal subcortical circuitry (Cummings 2003b).

Disturbances of the medial frontal-subcortical circuitry are manifested primarily by apathy, a common feature of conditions affecting the frontal lobe and related subcortical structures (Levy et al. 1996, 1998; Litvan et al. 1996, 1998a,b).

Several unique behaviors have been recognized in the patients with frontotemporal lobar degeneration and in occasional other tauopathies that are sufficiently distinctive that, when the phenotype is observed, it nearly invariably indicates the presence of the underlying proteotype abnormality. Among these features are compulsive behaviors with ritualistic and stereotype repetition of specific behavioral patterns, unusual alterations in dietary intake with an emphasis on sweets and carbohydrates, and the emergence of unusual artistic talents during the early and middle phases of the frontotemporal lobar degeneration process (Ames et al. 1994; Miller et al. 1998; Cummings 2003b).

Differences in the distribution of pathological changes among the tauopathies indicate that there is a differential cellular involvement even when the proteotype is similar. Genetic, environmental, and developmental influences likely account for these differences.

Phenotype of Parkinson's Disease and Other Alpha-Synucleinopathies

Pigmented brainstem neurons and glial cells are the cell populations that are most vulnerable to alpha-synuclein abnormalities. The spread of alpha-synuclein pathology from brain stem to limbic to neocortical structures provides an index of relative cellular vulnerability (Braak et al. 2003). Intracellular alpha-synuclein inclusions have been observed in glial cells in neurodegeneration with brain iron accumulation and in the multi-system atrophies (Trojanowski and Lee 2000).

The principal phenotypic feature shared among the synucleinopathies is parkinsonism. Akinetic rigid manifestations are evident across the spectrum of alpha-synuclein disorders. They may be mild in the case of dementia with Lewy bodies or severe and combined with tremor in the case of Parkinson's disease. Involvement of dopaminergic neurons in the brain stem and consequent dopaminergic denervation of basal ganglia structures results in executive dysfunction that overlaps phenotypically with the tauopathies. The severity of executive abnormalities observed in alpha-synuclein disorders is typically more mild than that observed in the tauopathies (Stout and Paulsen 2003). The aphasic type of language disorders observed in some frontotemporal dementia patients and in AD are not characteristic of Parkinson's disease or other alpha-synuclein disorders (Cohen 2003). A distinctive neuropsychological feature of Parkinson's disease dementia and dementia with Lewy bodies is the marked fluctuation in cognition (Ballard et al. 2001; Walker et al. 2000a,b; Ballard et al. 2002).

In addition to fluctuations, two other features assist in distinguishing alpha-synuclein disorders from other neurodegenerative diseases. These include visual hallucinations and rapid eye movement (REM) sleep behavior disorder. Fully formed complex visual hallucinations are common manifestations of both dementia with Lewy bodies and Parkinson's disease. While they typically follow the introduction of treatment with dopaminergic substances in patients with Parkinson's disease, they occur primarily in patients with a risk profile that includes cognitive impairment (Fenelon et al. 2000; Barnes and David 2001; Holroyd et al. 2001). A common feature of patients with alpha-synuclein disorders and prominent formed visual hallucinations is the presence of cognitive impairment.

REM sleep behavior disorder is characterized by the apparent acting out of dream content, with frequent shouting and violent behavior while the patient remains asleep. The syndrome occurs when muscle atonia and sleep paralysis loses its normal association with periods of REM (dreaming) sleep. REM sleep behavior disorder has a nearly unique relationship to alpha-synucleinopathies (Fantini et al. 2003; Ferman et al. 1999; Turner 2002; Ferini-Strambi et al. 2004; Gagnon et al. 2002, 2004; Comella et al. 1998).

Phenotype of AD and its Relation to Beta-Amyloid Protein

AD has a relatively stereotyped neuropsychological phenotype characterized by an amnestic type of memory disorder, aphasia beginning with a deficit in confrontation naming and progressing to transcortical sensory aphasia, prominent visuospatial deficits demonstrable by copying and drawing tasks and mild executive disturbances (Cummings 2000). Memory abnormalities differ from the retrieval deficit disorder with impaired recall and intact recognition that are typical of the tauopathies and alpha-synucleinopathies. The language abnormality of AD differs in character from the language changes that may characterize some tauopathies; AD produces a lexical, selection-type anomia, whereas semantic dementia features a semantic-type naming deficit and primary progressive aphasia is characterized by a word-production type anomia (Kertesz 1998; Hodges et al. 1998; Cummings 2000).

Neuropsychiatrically, AD has no specific distinctive features, but it is characterized by a plethora of neuropsychiatric manifestations, including apathy, agitation, depression, anxiety, and irritability with less evidence of disinhibition, delusions, and visual hallucinations (Mega et al. 1996).

AD is associated with a secondary tauopathy in the form of intracellular neurofibrillary tangles. When the tau-related changes are particularly abundant, they

Table 3. Distinctive phenotypic features of the major proteinopathies producing neurodegenerative disorders.

Protein	Disease	Phenotypic characteristics
Tau	Frontotemporal dementia Progressive supranuclear palsy Corticobasal degeneration	Executive dysfunction Compulsions Dietary changes Artistic creativity
Alpha-synuclein	Parkinson's disease Dementia with Lewy bodies	Parkinsonism Fluctuations in cognition Visual hallucinations REM* sleep behavior disorder
Amyloid protein	Alzheimer's disease	Amnestic-type memory disorder Lexical selection-type naming deficit

REM*-Rapid eye movement

occur disproportionally in the frontal lobe and produce a dysexecutive syndrome and behavioral abnormalities similar to those observed in frontotemporal dementia and other primary tauopathies Johnson et al. 1999; Tekin et al. 2001). Thus, when the tauopathy occurs within the context of a primary amyloid disorder, it retains its fundamental pathological distribution and associated phenotype.

Table 3 summarizes the distinctive phenotypic features of the major proteinopathies.

Implications for Treatment

Many shared pathological processes are now recognized among neurodegenerative disorders. Hereditary forms of neurodegenerative diseases tend to feature protein overproduction, whereas late-onset sporadic forms of the disease appear to result primarily from abnormal protein accumulation and reduced degradation. In both cases, cellular processes involved in protein metabolism are overwhelmed, typically leading to cellular toxicity and the accumulation of abnormal protein deposits. Deposition of beta-amyloid protein is primarily extracellular in the form of neuritic plaques; alpha-synuclein deposits take the form of intracellular Lewy bodies and tau accumulation may be in the form of neurofibrillary tangles or other tau-positive intracellular accumulations (Taylor et al. 2002; Lovestone and McLoughlin 2002; Ritchie and Lovestone 2002). Conformational changes in proteins with the associated neurotoxicity lead to a variety of secondary toxic and function abnormalities, including excitotoxicity, oxidative stress, mitochondrial injury, synaptic failure, inflammation, and altered metal homeostasis (Bossy-Wetzel et al. 2004; Scheff and Price 2003; Barnham et al. 2004; Masliah et al. 1996; Selkoe 2002; Kermer et al. 2004). This multiplicity of events offers a variety of molecular targets for therapies that may limit the changes in protein metabolism or ameliorate secondary consequences contributing to the cognitive, behavioral, and motoric compromises exhibited by patients with neurodegenerative disorders (Shoulson 1998; Trojanowski 2004). Torsin proteins reduce polyglutamine-induced protein aggregation in cellular models relevant to Huntington's disease and suppress alpha-synuclein aggregation in models relevant to Parkinson's disease (McLean et al. 2002; Caldwell et al. 2003). In preliminary studies, minocycline appears to have neuroprotective effects in mouse models of Huntington's disease (Chen et al. 2000) and amyotrophic lateral sclerosis (Zhu et al. 2002). In initial clinical trials, antioxidant therapy with co-enzyme Q slowed the progression of Parkinson's disease (Shults et al. 2002), and alpha-tocopherol and selegiline slowed the loss of functional symptoms in patients with AD (Sano et al. 1997). Memantine has been shown to improve cognition and function relative to placebo in AD, and the reduction in excitotoxicity mediated via n-methyl-d-aspartate (NMDA) receptor antagonist may be relevant to other diseases as well (Reisberg et al. 2003; Tariot et al. 2004). Evidence of inflammation has been observed in many neurodegenerative disorders, and treatment with anti-inflammatory compounds may delay the onset or slow the progression of cell dysfunction and death.

Earlier detection of patients at risk for neurodegenerative disease or evidencing the earliest manifestations, such as mild cognitive impairment, will allow initiation of therapy at a time when cellular injury is absent or mild. A better understanding of the processes that initiate protein misfolding may provide clues to early clinical detection through better recognition of at-risk phenotypes.

Biological markers that identify patients who are therapeutic candidates are also urgently needed. Proteomic analysis may allow detection of products of abnormal protein metabolism early in the disease process. High sensitivity analyses may also allow detection of evidence of secondary consequences (oxidation, inflammation, apoptosis) and allow implementation of process-specific therapeutic interventions. Amelioration of disease progress through intensive characterization of the phenotype remains the gold standard in determining the success of therapeutic intervention.

Acknowledgments

This work was supported by the Alzheimer's Disease Research Center Grant (AG16570) from the National Institute of Aging (NIA), an Alzheimer's Disease Research Center Grant from the state of California, the Sidell Kagan Foundation, and the Deane F. Johnson Center for Neurotherapeutics.

References

Ames D, Cummings JL, Wirshing WC, Quinn B, Mahler M (1994) Repetitive and compulsive behavior in frontal lobe degenerations. J Neuropsychiat Clin Neurosci 6:100–113.

Ballard C, O'Brien J, Gray A, Cormack F, Ayre G, Rowan E, Thompson P, Bucks R, Mckeith I, Walker M, Tovee M (2001) Attention and fluctuating attention in patients with dementia with Lewy bodies and Alzheimer disease. Arch Neurol 58:977–982.

Ballard CG, Aarsland D, McKeith I, O'Brien J, Gray A, Cormack F, Burn D, Cassidy T, Starfeldt R, Larsen JP, Brown R, Tovee M. (2002) Fluctuations in attention: PD dementia vs DLB with parkinsonism. Neurology 59:1714–1720.

Barnes J, David AS (2001) Visual hallucinations in Parkinson's disease: a review and phenomenological survey. J Neurol Neurosurg Psychiat 70:727–733.

Barnham KJ, Masters CL, Bush AI (2004) Neurodegenerative diseases and oxidative stress. Nature Rev Drug Discov 3:205–214.

Bird TD (1999) Clinical genetics of familial Alzheimer disease. In: Terry RD, Katzman R, Bick KL (eds) Alzheimer disease. 2nd ed. Philadelphia, PA: Lippincott Williams & Wilkins, pp. 57–66.

Blacker D, Tanzi R (2000) Genetic testing in the early diagnosis of Alzheimer's disease. In: Scinto LMF, Daffner KR (eds) Early diagnosis of Alzheimer's disease. Totowa, NJ: Humana Press, Inc., pp. 105–125.

Bossy-Wetzel E, Schwarzenbacher R, Lipton SA (2004) Molecular pathways to neurodegeneration. Nature Med 5:2–9.

Braak H, De Tredici K, Rub U, de Vos RA, Jansen Steur EN, Braak E (2003) Staging of brain pathology related to sporadic Parkinson's disease. Neurobiol Aging 24:197–211.

Caldwell GA, Cao S, Sexton EG, Gelwix CC, Bevel JP, Caldwell KA (2003) Suppression of polyglutamine-induced protein aggregation in Caenorhabditis elegans by torsin proteins. Human Mol Genet 12:307–319.

Chen M, Ona VO, Li M, Ferrante RJ, Fink KB, Zhu S, Bian J, Guo L, Farrell LA, Hersch SM, Hobbs W, Vonsattel JP, Cha JH, Friedlander RM (2000) Minocycline inhibits caspase-1 and caspase-3 expression and delays mortality in a transgenic mouse model of Huntington disease. Nature Med 6:797–801.

Cohen H (2003) Disorders of speech and language in Parkinson's disease. In: Bedard M-A (ed) Mental and behavioral dysfunction in movement disorders. Totowa, NJ: Humana Press, Inc., pp. 125–134.

Comella C, Nardine TM, Diedrich NJ, Stebbins GT (1998) Sleep-related violence, injury, and REM sleep behavior disorder in Parkinson's disease. Neurology 51:526–529.

Cummings JL (1993) Frontal-subcortical circuits and human behavior. Arch Neurol 50:873–880.

Cummings JL (2000) Cognitive and behavioral heterogeneity in Alzheimer's disease: seeking the neurological basis. Response to commentaries. Neurobiol Aging 21:845–861.

Cummings JL (2003a) Towards a molecular neuropsychiatry of neurodegenerative diseases. Ann Neurol 54:147–154.

Cummings JL (2003b) The Neuropsychiatry of Alzheimer's Disease and Related Dementia's. London: Martin Dunitz.

Cummings JL, Litvan I (2000) Neuropsychiatric aspects of corticobasal degeneration. Adv Neurol 82:147–152.

Dickson DW, Liu WK, Ksiezak-Reding H, Yen SH (2000) Neuropathologic and molecular considerations. In: Litvan I, Goetz CG, Lang AE (eds) Corticobasal degeneration. Advances in neurology. Philadelphia: Lippincott & Williams, pp. 9–27.

Fantini ML, Gagnon JF, Petit D, Rompre S, Decary A, Carrier J, Montplaisir J (2003) Slowing of electroencephalogram in rapid eye movement sleep behavior disorder. Ann Neurol 53:774–780.

Fenelon G, Mahieux F, Huon R, Ziegler M (2000) Hallucinations in Parkinson's disease. Prevalence, phenomenology and risk factors. Brain 123:733–745.

Ferini-Strambi L, Di Gioia MR, Castronovo V, Oldani A, Zucconi M, Cappa SF (2004) Neuropsychological assessment in idiopathic REM sleep behavior disorder (RBD). Does the idiopathic form of RBD really exist? Neurology 62:41–45.

Ferman TJ, Boeve BF, Smith GE, Silber MH, Kokmen E, Petersen RC, Ivnik RJ (1999) REM sleep behavior disorder and dementia: cognitive differences when compared with AD. Neurology 52:951–957.
Gagnon J-F, Bedard M-A, Fantini ML, Petit D, Panisset M, Rompre S, Carrier J, Montplaisir J (2002) REM sleep behavior disorder and REM sleep without atonia in Parkinson's disease. Neurology 59:585–589.
Gagnon JF, Fantini ML, Bedard MA, Petit D, Carrier J, Rompre S, Decary A, Panisset M, Montplaisir J (2004) Association between waking EEG slowing and REM sleep behavior disorder in PD without dementia. Neurology 62:401–406.
Hardy J, Gwinn-Hardy K (1998) Genetic classification of primary neurodegenerative disease. Science 282:1075–1079.
Hauw JJ, Daniel SE, Dickson D, Horoupian DS, Jellinger K, Lantos PL, McKee A, Tabaton M, Litvan I (1994) Preliminary NINDS neuropathologic criteria for Steele-Richardson-Olszewski syndrome (progressive supranuclear palsy). Neurology 44:2015–2019.
Hodges JR, Garrard P, Partterson K. Semantic dementia. (1998) In: Kertesz A, Munoz DG (eds) Pick's disease and Pick complex. New York: Wiley-Liss, Inc., pp. 83–104.
Holroyd S, Currie L, Wooten GF (2001) Prospective study of hallucinations and delusions in Parkinson's disease. J Neurol Neurosurg Psychiat 70:734–738.
Johnson JK, Head E, Kim R, Starr A, Cotman CW (1999) Clinical and pathological evidence for a frontal variant of Alzheimer's disease. Arch Neurol 56:1233–1239.
Kermer P, Liman J, Weishaupt JH, Bähr M (2004) Neuronal apoptosis in neurodegenerative diseases: from basic research to clinical application. Neurodegen Dis 1:9–19.
Kertesz A (1998) Primary progressive aphasia. In: Kertesz A, Munoz DG (eds) Pick's disease and Pick complex. New York: Wiley-Liss, Inc., pp. 69–81.
Levy ML, Miller BL, Cummings JL, Fairbanks LA, Craig A (1996) Alzheimer disease and frontotemporal dementias: behavioral distinctions. Arch Neurol 53:687–690.
Levy ML, Cummings JL, Fairbanks LA, Masterman D, Miller BL, Craig AH, Paulsen JS, Litvan I (1998) Apathy is not depression. J Neuropsychiat Clin Neurosci 10:314–319.
Litvan I, Mega MS, Cummings JL, Fairbanks L (1996) Neuropsychiatric aspects of progressive supranuclear palsy. Neurology 47:1184–1189.
Litvan I, Paulsen JS, Mega MS, Cummings JL (1998) Neuropsychiatric assessment of patients with hyperkinetic and hypokinetic Mov Disord Arch Neurol 55:1313–1319.
Litvan I, Cummings JL, Mega M (1998) Neuropsychiatric features of corticobasal degeneration. J Neurol Neurosurg Psychiat 65:717–721.
Lovestone S, McLoughlin DM (2002) Protein aggregates and dementia: is there a common toxicity? J Neurol Neurosurg Psychiat 72:152–161.
Masliah E, Alford M, DeTeresa R, Mallory M, Hansen L (1996) Deficient glutamate transport is associated with neurodegeneration in Alzheimer's disease. Ann Neurol 40:759–766.
McLean PJ, Kawamata H, Shariff S, Hewett J, Sharma N, Ueda K, Breakefield XO, Hyman BT (2002) TorsinA and heat shock proteins act as molecular chaperones: suppression of alpha-synuclein aggregation. J Neurochem 83:846–854.
Mega MS, Cummings JL, Fiorello T, Gornbein J (1996) The spectrum of behavioral changes in Alzheimer's disease. Neurology 46:130–135.
Miller B, Boone K, Mishkin F, Swartz JR, Koras N, Kushii J (1998) Clinical and neuropsychological features of frontotemporal dementia. In: Kertesz A, Munoz DG (eds) Pick's disease and Pick complex. New York: Wiley-Liss, Inc., pp. 23–32.
Munoz DG (1998) The pathology of Pick complex. In: Kertesz A, Munoz DG (eds) Clinical and neuropsychological features of frontotemporal dementia. New York, Wiley-Liss, Inc., pp. 211–241.
Reisberg B, Doody R, Stoffler A, Schmitt F, Ferris S, Mobius HJ; Memantine Study Group (2003) Memantine in moderate-to-severe Alzheimer's disease. New Engl J Med 348:1333–1341.
Ritchie K, Lovestone S (2002) The dementias. Lancet 360:1759–1766.

Ross CA, Poirier MA (2004) Protein aggregation and neurodegenerative disease. Nature Med 5:10–17.
Sanchez-Ramos J, Ortoll R, Paulson G (1996) Visual hallucinations associated with Parkinson disease. Arch Neurol 53:1265–1268.
Sano M, Ernesto C, Thomas RG, Klauber MR, Schafer K, Grundman M, Woodbury P, Growdon J, Cotman CW, Pfeiffer E, Schneider LS, Thal LJ (1997) A controlled trial of selegiline, alpha-tocopherol, or both as treatment for Alzheimer's disease. New Engl J Med 336:1216–1222.
Scheff SW, Price DA (2003) Synaptic pathology in Alzheimer's disease: a review of ultrastructural studies. Neurobiol Aging 24:1029–1046.
Selkoe DJ (2002) Alzheimer's disease is a synaptic failure. Science 298:789–791.
Shoulson I (1998) Experimental therapeutics of neurodegenerative disorders: unmet needs. Science 282:1072–1074.
Shults CW, Oakes D, Kieburtz K, Beal MF, Haas R, Plumb S, Juncos JL, Nutt J, Shoulson I, Carter J, Kompoliti K, Perlmutter JS, Reich S, Stern M, Watts RL,Kurlan R, Molho E, Harrison M, Lew M; Parkinson Study Group (2002) Effects of coenzyme Q10 in early Parkinson disease: evidence of slowing of the functional decline. Arch Neurol 59:1541–1550.
Soto C (2003) Unfolding the role of protein misfolding in neurodegenerative diseases. Nat Rev Neurosci 4:49–60.
Stout JC, Paulsen JS (2003) Assessing cognition in movement disorders. In: Bedard M-A (ed) Mental and behavioral dysfunction in movement disorders. Totowa, NJ: Humana Press, Inc., pp. 85–99.
Tariot PN, Farlow MR, Grossberg GT, Graham SM, McDonald S, Gergel I, Memantine Study Group (2004) Memantine treatment in patients with moderate to severe Alzheimer disease already receiving donepezil: a randomized controlled trial. JAMA 291:317–324.
Taylor JP, Hardy J, Fischbeck KH (2002) Toxic proteins in neurodegenerative disease. Science 296:1991–1995.
Tekin S, Mega MS, Masterman DL, Chow T, Garakian J, Vinters HV, Cummings JL (2001) Orbitofrontal and anterior cingulate cortex: neurofibrillary tangle burden is associated with agitation in Alzheimer's disease. Ann Neurol 49:355–361.
Trojanowski JQ (2004) Protein mis-folding emerges as a "drugable" target for discovery of novel therapies for neuropsychiatric diseases of aging. Am J Geriatr Psychiat 12:134–135.
Turner RS (2002) Idiopathic rapid eye movement sleep behavior disorder is a harbinger of dementia with Lewy bodies. J Geriatr Psychiat Neurol 15:195–199.
Trojanowski JQ, Lee VM (2000) "Fatal attractions" of proteins. A comprehensive hypothetical mechanism underlying Alzheimer's disease and other neurodegenerative disorders. Ann NY Acad Sci 924:62–67.
Walker MP, Ayre GA, Cummings JL, Wesnes K, McKeith IG, O'Brien JT, Ballard CG (2000) Quantifying fluctuation in dementia with Lewy bodies, Alzheimer's disease, and vascular dementia. Neurology 54:1616–1624.
Walker MP, Ayre GA, Cummings J, Wesnes K, McKeith IG, O'Brien JT, Ballard CG (2000) The clinician assessment of fluctuation and the one day fluctuation assessment scale. Two methods to assess fluctuating confusion in dementia. Brit J Psychiat 177:252–256.
Zhu S, Stavrovskaya IG, Drozda M, Kim BYS, Ona V, Mingwei Li, Satinder Sarang, Liu AS, Hartley DM, Wu DC, Gullans S, Ferrante RJ, Przedborski S, Kristal BS, Friedlander RM (2002) Minocycline inhibits cytochrome c release and delays progression of amyotrophic lateral sclerosis in mice. Nature 417:74–78.

Towards a Molecular Classification of Neurodegenerative Disease

John Hardy[1]

Traditionally, diseases have been classified as clinicopathological entities: a patient reports a distinctive set of symptoms, a clinician sees a distinctive set of progressive signs and symptoms and, on pathological exam, the pathologist reports lesions in brain regions that underpin these signs and symptoms. Nearly all neuropharmacologic interventions are based on the pathoanatomy of the lesions. This approach to disease has been extremely successful, and Western treatment of disease is largely based on it. However, this approach to disease has been palliative and there has been no knowledge of the causes of the damage.

Molecular genetic analysis is having an impact on this approach to disease in two ways. First, it is offering a window on the etiology and pathogenesis of disease, making clear how diseases may be initiated. Second, it is showing that the boundaries of diseases are not where they might have been expected to be (Hardy and Gwinn-Hardy 1999). However, it has not yet had significant impact on clinical practice beyond genetic counseling. It is to be hoped that treatment strategies based on genetic knowledge will soon be developed.

The purpose of this chapter is threefold:
1) to discuss how genetic analysis has allowed the dissection of disease etiology,
2) to discuss how this has impacted on disease definition, and
3) to suggest how changing diagnostic and therapeutic practices may result from genetics-based treatments.

Molecular Genetic Dissection Of Disease Etiology

The success of molecular genetic analysis in identifying the underlying defects in Mendelian disease is well known; examples include prion mutations in hereditary spongiform encephalopathies (Owen et al. 1989), APP (Goate et al. 1991) and presenilin mutations in autosomal dominant Alzheimer's disease (Sherrington et al. 1995; Rogaev et al. 1995; Levey-Lahad et al. 1995), tau in Pick's disease (frontal temporal dementia with parkinsonism linked to chromosome 17: FTDP-17) (Poorkaj et al. 1998; Hutton et al. 1998), α-synuclein in Parkinson's disease (Polymeropoulos et al. 1997) and multiple other genes in other parkinsonian syn-

[1] Laboratory of Neurogenetics, National Institute on Aging, National Institutes of Health, Building 35, Room 1A1015, MSC 3707, Bethesda, MD 20892, USA

Cummings et al.
Genotype – Protéotype – Phenotype
Relationships in Neurodegenerative Diseases

dromes (Kitada et al. 1998; Bonifati et al. 2003; Valente et al. 2004), superoxide dismutase mutations in ALS (Rosen et al. 1993) and multiple genes in other motor neuron disease syndromes (Yang et al. 2001; Hadano et al. 2001; Hafezparast et al. 2003) as well as multiple genes in ataxic syndromes (Pulst 2003). In each of these cases, the identification of the pathogenic mutations clearly marks the initiating event in disease pathogenesis. Two issues are more complex, however:

1) if there are mutations in more than one gene that lead to disease, to what extent can one suppose they mark a single pathway to cell death?
2) to what extent is it appropriate to infer that the pathogenetic mechanisms from the rare Mendelian varieties of disease are involved in the more common sporadic forms of disease?

One pathway to cell death in different Mendelian forms of disease?

In Alzheimer's disease, parkinsonian syndromes, motor neuron disease and ataxic syndromes, there are mutations in multiple genes that can lead to disease, and clearly it is important to determine whether these mutations mark a single pathway to neurodegenerative disease or whether each acts in an entirely independent way.

The most data are available for Alzheimer's disease, in which the Mendelian mutations are to the substrate and the enzyme responsible for liberating Aβ from APP by γ-secretase cleavage (Wolfe et al. 1999). Clearly, in this case, it is parsimonious to suggest that these mutations lead to disease through the same pathway, starting with Aβ and leading through the other aspects of the pathology: tangles, Lewy bodies and cell death (Hardy 1997). However, it is likely that the pathways to cell death overlap with those of Pick's disease and Parkinson's disease, since cases of Alzheimer's disease with either APP or presenilin mutations develop tangles and Lewy bodies (Lantos et al. 1992; Lippa et al. 2001) and APP transgenes potentiate both tau (Lewis et al. 2000, 2001) and synuclein pathology in mice (Masliah et al. 2000, 2001) with the relevant transgenes. Thus, in Alzheimer's disease, it seems that Aβ can initiate two (at least) pathways to cell death: one, through tau and tangles, that resembles the cell death pathway initiated by tau mutations (Hardy et al. 1998), and one through α-synuclein and Lewy bodies that resembles the cell death pathway initiated by α-synuclein mutations (Hardy 1994).

The situation in parkinsonian disorders is more complex. Autosomal dominant mutations in α-synuclein lead to a disease that often resembles idiopathic Parkinson's disease (Golbe et al. 1999) but sometimes resembles Lewy Body Dementia (Muenter et al. 1998; Zarranz et al. 2004); however, the disease always has a pathology that is similar to, though often more extreme than, typical idiopathic disease (Gwinn-Hardy et al. 2000b).

Mutations in parkin cause a disease that usually has an early onset (Kitada et al. 1998) and a very slow course, is recessive in inheritance and has no Lewy body pathology (Hattori and Mizuno 2004). However, exceptions to all these statements have been reported; associations with late onset and typical course have been reported (Lucking et al. 2000), as have kindreds with autosomal dominant disease (Klein et al. 2000) and Lewy body pathology. Parkin is a E3 ubiquitin ligase

(Shimura et al. 2000), and one of its functions seems to be in protein degradation, although there may be other functions. The one report of a direct interaction between parkin and α-synuclein has not been replicated (Shimura et al. 2001).

Mutations in DJ-1 (Bonifati et al. 2003) and PINK1 (Valente et al. 2004) seem to lead to phenotypes that are similar to typical parkin encoded disease:, that is, early onset, recessive inheritance with a slow course, and good response to dopaminergic medication, but no neuropathological examinations have been reported yet. DJ-1 seems to be involved in signaling to the mitochondria when there are conditions of oxidative stress (Canet-Aviles et al. 2004) and PINK1 is a mitochondrial kinase (Valente et al. 2004).

Thus, with hereditary parkinsonisms, it is very difficult to be certain whether there is one or more pathways to cell death. The main problem is that there are few or no reports of neuropathological examinations of mutation carriers and no definitive molecular experiments linking the cognate genes to a single pathway.

With respect to motor neuron diseases, the nomenclature is even messier than with the parkinsonisms, and the clinical phenotypes of ALS overlap with that of Charcot Marie Tooth disease and hereditary motor neuronopathies (De Jonghe et al. 2000). The major locus, ALS1, is superoxide dismutase (SOD1; usually autosomal dominant; Rosen et al. 1993), and others include alsin (a GTPAse regulator of unknown function; Yang et al. 2001; Hadano et al. 2001) and dynactin (dominant; Puls et al. 2003), part of the retrograde transport system. There is no molecular evidence of interactions between these molecules, and the dearth of pathological reports concerning the pathology of such cases makes it very difficult to assess whether these proteins and others that are implicated in the loss of motor neurons mark one or many pathways.

The majority of ataxias are polyglutamine expansion disorders and they, together with Huntington's disease, show intranuclear inclusions of polyglutamine (Davies et al. 1999). With the similarity of the mutations and the pathology, as well as the usual overlap of the clinical symptoms (but see below), one has to suppose that disease mechanisms are shared. One interesting possible exception is SCA6, which involves polyglutamine expansion in a calcium channel (Zhuchenko et al. 1997), other mutations in which cause episodic ataxia (Ophoff et al. 1996), leaving open the possibility that, with this disease if not the others, there may be other routes to cell death in play.

Is the pathogenic mechanism implicated in Mendelian forms of disease relevant to sporadic disease?

In general, the sporadic forms of all the diseases are more common than the ones with simple modes of inheritance: in Alzheimer's disease and in Parkinson's, for example, less than 1% of the "typical" cases have clear modes of inheritance. Clearly, it is of great importance to have some insight into whether the common forms of disease are related, in terms of disease mechanism, to the hereditary forms.

On this issue, the most data are available for prion diseases; it is clear that the prion protein is central and of initiating importance in all cases of the disease (Prusiner 1998), and prion protein is the major component of the infectious agent in these diseases. Prion genotype influences susceptibility to infectious (Collinge et al. 1991) and sporadic disease (Palmer et al. 1991), and genetic evidence suggests that expression levels of the prion protein are important in determining susceptibility to sporadic disease (Mead et al. 2001). However, puzzlingly, many of the mutant forms of protein do not lend themselves to infectivity, suggesting there may well be some subtle differences in pathogenesis.

In Alzheimer's disease, the argument about whether one can extrapolate from the Mendelian forms of the disease to the more common, sporadic forms of disease has a long history. However, there are two moderately strong arguments that this extrapolation is justified. First, the only established genetic risk factor for typical late onset disease is apolipoprotein E (Corder et al. 1993): the precise role of apoE in Alzheimer's disease remains unclear but it does seem that it is involved in Aβ metabolism (Bales et al. 1999). Second, genetic analysis of sibpairs with late onset Alzheimer's disease suggests two loci – one on chromosome 10 (Myers et al. 2000; Erteken-Taner et al. 2000) and the other at the APP gene itself on chromosome 21 (Wavrant-DeVrieze et al. 1999) – that influence Aβ metabolism. Of course, the occurrence of Alzheimer's disease in Down syndrome (Olson and Shaw 1969; Prasher et al. 1998) implicates overproduction of Aβ as a risk factor for the disease in general.

In the tau diseases, the position is very interesting: tau mutations cause a wide range of phenotypes ranging from parkinsonism to dementia (Foster et al. 1997), with very variable tau pathology (Spillantini et al. 1998). Among the sporadic diseases, it is clear that the tau haplotype is a risk factor locus for the diseases, including progressive supranuclear palsy (Baker et al. 1999), corticobasal degeneration (Houlden et al. 2000) and argyrophilic grain disease (Togo et al. 2002). Thus, in these diseases at least, the "diagnoses" are different between the hereditary and sporadic diseases, but in some cases at least, the pathogenic mechanisms are similar and are presumably shared with some of the pathogenic mechanisms of cell loss in Alzheimer's disease.

The situation in Parkinson's disease is perhaps the most clear with respect to α-synuclein. Genetic variability at the α-synuclein locus contributes to the risk of sporadic disease (Farrer et al. 2001), with the high expressing haplotypes being associated with greatest risk (Chiba-Falek and Nussbaum 2001). This observation is concordant with the observation of individuals with triplication and duplications of the synuclein locus having autosomal dominant disease (Singleton et al. 2003). The roles of parkin, DJ-1 and PINK1 in idiopathic disease remain unclear.

For ALS and for the ataxias, the lack of detailed histopathology in most cases of disease and the lack of genetic association studies mean that there is little that can be adduced concerning the relationship between Mendelian and sporadic disease.

In conclusion, therefore, the bulk of the evidence suggests that, certainly in prion disease, Alzheimer's disease, and the tau and synuclein diseases, genetic variability at the autosomal dominant loci contributes to the risk of sporadic disease, suggesting that the sporadic diseases and the Mendelian diseases share pathogenic mechanisms (Singleton et al. 2004). This finding is an enormously important consideration for the development of treatments because these, in large part, are validated on genetic models of disease.

The impact of molecular genetics on disease classifications

Whenever a gene is discovered for disease, it always leads to the realization that the disease has a different and broader spectrum than was suspected. Again, prion diseases led the way, first, with the realization that "spongiform encephalopathies" (as the diseases were then named) included cases without spongiform change (Collinge et al. 1990) and, later, with the realization that fatal familial insomnia was a prion disease (Medori et al. 1992). In these cases, while we do not fully understand the reasons for the phenotypic variability in disease expression, there are always five options that need to be considered:

1) genetic variability in cis in other words, the mutation itself or something else, such as the variability in the promoter influences phenotype
2) genetic variability on the trans allele: in this case, variability on the patients' other prion allele influences phenotype
3) variability elsewhere in the genome influences phenotype
4) environmental influences on phenotype
5) stochastic effects influence phenotype

In the prion diseases, it is clear that cis effects are important: the mutant allele haplotype clearly influences the phenotype with, for example, fatal familial insomnia and Creuzfeldt Jakob disease, because it is caused by the same mutation on different backgrounds (Goldfarb et al. 1992). In addition, trans alleles can often affect the age of onset of prion disease with codon 129 homozygosity, but not always, because they are associated with earlier onset ages (Owen et al. 1990). There are no data yet concerning variability elsewhere in the genome affecting the disease phenotype, but recent mouse data suggest that there are prion disease-susceptibility alleles (Lloyd et al. 2001). It will be interesting to test these as phenotype modifiers when they are identified.

In Alzheimer's disease, the phenotype is extraordinarily constant. However, even here the precise mutation leads to different ages of onset, probably determined partly by the size of its cellular effects on Aβ production (Murayama et al. 1999). In addition, it is clear that apoE genotype affects onset ages in both APP and presenilin families (Houlden et al. 1993; Pastor et al. 2003). There are one definite and two possible exceptions to this homogeneity of phenotype of Alzheimer's disease. The definite exception is the occurrence of the spastic paraparesis/cotton wool plaque variant of Alzheimer's disease (Crook et al. 1998), which is caused by those presenilin mutations that, in cell systems, have particularly large effects on APP processing (Houlden et al. 2000). In addition, it seems that some presenilin mutations present as a frontal dementia (Tang-Wai et al. 2002; Dermaut et al. 2004). The possible and potentially most important exception is the suggestion that the presentation of Alzheimer's disease in some of those of African background can be more "frontal," with personality disorder being a presenting sign

(Rippon et al. 2003), although this finding is not true in all families of African descent (Heckmann et al. 2004).

In tau diseases, the range of the phenotype, both pathological and clinical, is extraordinarily large, with purely dementing and purely parkinsonian phenotypes being at the clinical extremes (Foster et al. 1997). The precise mutation clearly affects the pathological phenotype to a great extent (Hutton et al. 1998; Spillantini et al. 1998), but the cis haplotypic background and the trans allele haplotype may also play a role, with the suggestions that H1 homozygotes (Baker et al. 1998) tend to have a parkinsonian phenotype and H2 alleles may be more likely to lead to a dementing phenotype (Walker et al. 2002; Woodruff et al. 2004).

In α-synuclein diseases, there has been an ongoing and previously inconclusive debate about whether Parkinson's disease and Lewy body dementia are one disease or distinct entities: they are characterized by similar Lewy body pathology but with different, though overlapping, distributions (Gibb 1989; McKeith et al. 1996). Genetic analysis has shown that these phenotypes can occur in the same families, showing clearly that they involve the same pathogenic process (Zarranz et al. 2004; Muenter et al. 1998). The variables influencing this phenotypic spread are not clear, but the apolipoprotein E genotype may contribute to this variability (Lippa et al. 1995). Apolipoprotein E4 alleles predispose to Aβ pathology (Bales et al. 1999) and, in mice, Aβ pathology potentiates synuclein pathology (Masliah et al. 2001), so such an interaction is mechanistically plausible.

Finally, in the ataxias, an interesting observation has been that SCA2 and SCA3 in European populations are associated with ataxia (Pulst 2003), but in Chinese populations, SCA2 mutations, and in African populations, SCA3 mutations are often associated with parkinsonian phenotypes (Gwinn-Hardy et al. 2000, 2001). These observations relate to the distributions of the phenotypes and are not absolutes (Furtado et al. 2004; Subramony et al. 2002), but they most likely relate to the influence of genetic variability elsewhere in the genome (Hardy et al. 2003).

Thus, the effects of cis variability on varying disease phenotype have been frequently established: trans allele effects have been demonstrated in prion and possibly in tau diseases; the effects of genetic variability on disease phenotype have been firmly demonstrated for the apolipoprotein E locus in Alzheimer's disease and may also be important in synuclein diseases; and genome-wide allelic effects are likely to be important in the observation of racial differences of disease expressivity. Environmental and stochastic influences have never been systematically examined but probably also play a role in determining disease outcome: indeed, if one considers drug treatment as an "environmental influence," it is our hope that we can influence disease outcome through a changing environment.

Changing diagnostic practices will result from genetics-based treatments

Presently, most treatments are neurotransmitter based and their prescription is largely determined by diagnosis. Diagnosis is dependent on the pathoanatomy of the disease, which underpins the transmitter deficit. Of course, this is a simplistic view and fuzzy logic operates at all points in this chain: in practice, clinicians use whatever "works" in their patient. However, to effectively use therapies based on genetic knowledge, which are aimed at pathogenesis rather than at palliation, clinicians will need to be able to recognize etiologies: "Parkinson's disease" caused by a SCA2 repeat expansion will respond differently to treatment than Parkinson's disease: Lewy Body Dementia and Parkinson's disease will, however, probably respond to the same treatments. Progressive supranuclear palsy and Alzheimer's disease may respond to the same drugs if they target tau-based neurodegeneration, but only Alzheimer's disease will respond to therapies aimed at Aβ. Clearly, the major problem ahead of us is developing therapies guided by genetic knowledge (Hardy 2004), but a subsidiary issue is making sure that our diagnostic practices ensure the right patients receive each drug. While it may seem as if this is just a theoretical concern, it is possible that the reason a recent therapeutic trial for ALS failed in humans (Groeneveld et al. 2003) but worked in SOD1 transgenic mice (Klivenyi et al. 1999) was because it is not yet clear whether idiopathic ALS is pathogenically related to SOD1-encoded disease. Thus, determining pathogenesis on a patient-by-patient basis will be increasingly important.

References

Baker M, Litvan I, Houlden H, Adamson J, Dickson D, Perez-Tur J, Hardy J, Lynch T, Bigio E, Hutton M (1999) Association of an extended haplotype in the tau gene with progressive supranuclear palsy. Human Mol Genet 8:711–715.

Bales KR, Verina T, Cummins DJ, Du Y, Dodel RC, Saura J, Fishman CE, DeLong CA, Piccardo P, Petegnief V, Ghetti B, Paul SM (1999) Apolipoprotein E is essential for amyloid deposition in the APP(V717F) transgenic mouse model of Alzheimer's disease. Proc Natl Acad Sci USA 96:15233–15238

Bonifati V, Rizzu P, van Baren MJ, Schaap O, Breedveld GJ, Krieger E, Dekker MC, Squitieri F, Ibanez P, Joosse M, van Dongen JW, Vanacore N, van Swieten JC, Brice A, Meco G, van Duijn CM, Oostra BA, Heutink P (2003) Mutations in the DJ-1 gene associated with autosomal recessive early-onset parkinsonism. Science 299:256–259

Canet-Aviles RM, Wilson MA, Miller DW, Ahmad R, McLendon C, Bandyopadhyay S, Baptista MJ, Ringe D, Petsko GA, Cookson MR (2004) The Parkinson's disease protein DJ-1 is neuroprotective due to cysteine-sulfinic acid-driven mitochondrial localization. Proc Natl Acad Sci USA 101:9103–9108.

Chiba-Falek O, Nussbaum RL (2001) Effect of allelic variation at the NACP-Rep1 repeat upstream of the alpha-synuclein gene (SNCA) on transcription in a cell culture luciferase reporter system. Human Mol Genet 10:3101–3109

Collinge J, Owen F, Poulter M, Leach M, Crow TJ, Rossor MN, Hardy J, Mullan MJ, Janota I, Lantos PL (1990) Prion dementia without characteristic pathology. Lancet 336:7–9.

Collinge J, Palmer MS, Dryden AJ (1991) Genetic predisposition to iatrogenic Creutzfeldt-Jakob disease. Lancet 337:1441–1442.

Corder EH, Saunders AM, Strittmatter WJ, Schmechel DE, Gaskell PC, Small GW, Roses AD, Haines JL, Pericak-Vance MA (1993) Gene dose of apolipoprotein E type 4 allele and the risk of Alzheimer's disease in late onset families. Science 261:921–923.

Crook R, Verkkoniemi A, Perez-Tur J, Mehta N, Baker M, Houlden H, Farrer M, Hutton M, Lincoln S, Hardy J, Gwinn K, Somer M, Paetau A, Kalimo H, Ylikoski R, Poyhonen M, Kucera S, Haltia M (1998) A variant of Alzheimer's disease with spastic paraparesis and unusual plaques due to deletion of exon 9 of presenilin 1. Nature Med 4:452–455.

Davies SW, Sathasivam K, Hobbs C, Doherty P, Mangiarini L, Scherzinger E, Wanker EE, Bates GP (1999) Detection of polyglutamine aggregation in mouse models. Methods Enzymol 309:687–701.
De Jonghe P, Auer-Grumbach M, Irobi J, Wagner K, Plecko B, Kennerson M, Zhu D, De Vriendt E, Van Gerwen V, Nicholson G, Hartung HP, Timmerman V (2002) Autosomal dominant juvenile amyotrophic lateral sclerosis and distal hereditary motor neuronopathy with pyramidal tract signs: synonyms for the same disorder? Brain 125:1320–1325.
Dermaut B, Kumar-Singh S, Engelborghs S, Theuns J, Rademakers R, Saerens J, Pickut BA, Peeters K, van den Broeck M, Vennekens K, Claes S, Cruts M, Cras P, Martin JJ, Van Broeckhoven C, De Deyn PP (2004) A novel presenilin 1 mutation associated with Pick's disease but not beta-amyloid plaques. Ann Neurol 55:617–626.
Ertekin-Taner N, Graff-Radford N, Younkin LH, Eckman C, Baker M, Adamson J, Ronald J, Blangero J, Hutton M, Younkin SG (2000) Linkage of plasma Abeta42 to a quantitative locus on chromosome 10 in late-onset Alzheimer's disease pedigrees. Science 290:2303–2304.
Farrer M, Maraganore DM, Lockhart P, Singleton A, Lesnick TG, de Andrade M, West A, de Silva R, Hardy J, Hernandez D (2001) alpha-Synuclein gene haplotypes are associated with Parkinson's disease. Human Mol Genet 10:1847–1851
Foster NL, Wilhelmsen K, Sima AA, Jones MZ, D'Amato CJ, Gilman S (1997) Frontotemporal dementia and parkinsonism linked to chromosome 17: a consensus conference. Conference Participants. Ann Neurol 41:706–715.
Furtado S, Payami H, Lockhart PJ, Hanson M, Nutt JG, Singleton AA, Singleton A, Bower J, Utti RJ, Bird TD, de la Fuente-Fernandez R, Tsuboi Y, Klimek ML, Suchowersky O, Hardy J, Calne DB, Wszolek ZK, Farrer M, Gwinn-Hardy K, Stoessl AJ (2004) Profile of families with parkinsonism-predominant spinocerebellar ataxia type 2 (SCA2). Mov Disord 19:622–629.
Gibb WR (1989) Dementia and Parkinson's disease. Br J Psychiat154:596–614.
Goate A, Chartier-Harlin MC, Mullan M, Brown J, Crawford F, Fidani L, Giuffra L, Haynes A, Irving N, James L, Mant R, Newton P, Rooke K, Roques P, Talbot C, Pericak-Vance M, Roses A, Williamson R, Rossor M, Owen M, Hardy J (1991) Segregation of a missense mutation in the amyloid precursor protein gene with familial Alzheimer's disease. Nature 349:704–706.
Golbe LI, Di Iorio G, Lazzarini A, Vieregge P, Gershanik OS, Bonavita V, Duvoisin RC (1999) The Contursi kindred, a large family with autosomal dominant Parkinson's disease: implications of clinical and molecular studies. Adv Neurol 80:165–170.
Goldfarb LG, Petersen RB, Tabaton M, Brown P, LeBlanc AC, Montagna P, Cortelli P, Julien J, Vital C, Pendelbury WW, Haltia M; Willis PR; Hauw JJ; McKeever PE; Monari L; Schrank B; Swergold GD; Autilio-Gambetti L; Carleton Gajdusek D; Lugaresi E; Gambetti P (1992) Fatal familial insomnia and familial Creutzfeldt-Jakob disease: disease phenotype determined by a DNA polymorphism. Science 258:806–808.
Groeneveld GJ, Veldink JH, van der Tweel I, Kalmijn S, Beijer C, de Visser M, Wokke JH, Franssen H, van den Berg LH (2003) A randomized sequential trial of creatine in amyotrophic lateral sclerosis. Ann Neurol 53:437–445.
Gwinn-Hardy K, Chen JY, Liu HC, Liu TY, Boss M, Seltzer W, Adam A, Singleton A, Koroshetz W, Waters C, Hardy J, Farrer M (2000a) Spinocerebellar ataxia type 2 with parkinsonism in ethnic Chinese. Neurology 55:800–805.
Gwinn-Hardy K, Mehta ND, Farrer M, Maraganore D, Muenter M, Yen SH, Hardy J, Dickson DW (2000b) Distinctive neuropathology revealed by alpha-synuclein antibodies in hereditary parkinsonism and dementia linked to chromosome 4p. Acta Neuropathol (Berl) 99:663–672.
Gwinn-Hardy K, Singleton A, O'Suilleabhain P, Boss M, Nicholl D, Adam A, Hussey J, Critchley P, Hardy J, Farrer M (2001) Spinocerebellar ataxia type 3 phenotypically resembling parkinson disease in a black family. Arch Neurol 58:296–299.
Hadano S, Hand CK, Osuga H, Yanagisawa Y, Otomo A, Devon RS, Miyamoto N, Showguchi-Miyata J, Okada Y, Singaraja R, Figlewicz DA, Kwiatkowski T, Hosler BA, Sagie T, Skaug J, Nasir J, Brown RH Jr, Scherer SW, Rouleau GA, Hayden MR, Ikeda JE (2001) A gene encod-

ing a putative GTPase regulator is mutated in familial amyotrophic lateral sclerosis 2. Nature Genet 29:166–173.

Hafezparast M, Klocke R, Ruhrberg C, Marquardt A, Ahmad-Annuar A, Bowen S, Lalli G, Witherden AS, Hummerich H, Nicholson S, Morgan PJ, Oozageer R, Priestley JV, Averill S, King VR, Ball S, Peters J, Toda T, Yamamoto A, Hiraoka Y, Augustin M, Korthaus D, Wattler S, Wabnitz P, Dickneite C, Lampel S, Boehme F, Peraus G, Popp A, Rudelius M, Schlegel J, Fuchs H, Hrabe de Angelis M, Schiavo G, Shima DT, Russ AP, Stumm G, Martin JE, Fisher EM (2003) Mutations in dynein link motor neuron degeneration to defects in retrograde transport. Science 300:808–812.

Hardy J (1994) Lewy bodies in Alzheimer's disease in which the primary lesion is a mutation in the amyloid precursor protein. Neurosci Lett 180:290–291

Hardy J (1997) Amyloid, the presenilins and Alzheimer's disease. Trends Neurosci 20:154–159.

Hardy J (2004) Toward Alzheimer therapies based on genetic knowledge. Annu Rev Med 55:15–25

Hardy J, Gwinn-Hardy K (1999) Neurodegenerative disease: a different view of diagnosis. Mol Med Today 5:514–517.

Hardy J, Duff K, Hardy KG, Perez-Tur J, Hutton M. Genetic dissection of Alzheimer's disease and related dementias: amyloid and its relationship to tau. Nature Neurosci 1:355–358.

Hardy J, Singleton A, Gwinn-Hardy K (2003) Ethnic differences and disease phenotypes. Science 300:739–740.

Hattori N, Mizuno Y (2004) Pathogenetic mechanisms of parkin in Parkinson's disease. Lancet 364:722–724.

Heckmann JM, Low WC, de Villiers C, Rutherfoord S, Vorster A, Rao H, Morris CM, Ramesar RS, Kalaria RN (2004) Novel presenilin 1 mutation with profound neurofibrillary pathology in an indigenous Southern African family with early-onset Alzheimer's disease. Brain 127(Pt 1):133–142.

Houlden H, Baker M, McGowan E, Lewis P, Hutton M, Crook R, Wood NW, Kumar-Singh S, Geddes J, Swash M, Scaravilli F, Holton JL, Lashley T, Tomita T, Hashimoto T, Verkkoniemi A, Kalimo H, Somer M, Paetau A, Martin JJ, Van Broeckhoven C, Golde T, Hardy J, Haltia M, Revesz T (2000) Variant Alzheimer's disease with spastic paraparesis and cotton wool plaques is caused by PS-1 mutations that lead to exceptionally high amyloid-beta concentrations. Ann Neurol 48:806–808.

Houlden H, Collinge J, Kennedy A (1993) ApoE genotype and Alzheimer's disease. Lancet 342: 737–738.

Hutton M, Lendon CL, Rizzu P, Baker M, Froelich S, Houlden H, Pickering-Brown S, Chakraverty S, Isaacs A, Grover A, Hackett J, Adamson J, Lincoln S, Dickson D, Davies P, Petersen RC, Stevens M, de Graaff E, Wauters E, van Baren J, Hillebrand M, Joosse M, Kwon JM, Nowotny P, Kueiche L, Norton J, Morris JC, Reed LA, Trojanowski J, Basun H, Lannfelt L, Neystat M, Fahn S, Dark F, Tannenberg T, Dodd PR, Hayward N, Kwok JBJ, Schofield PR, Andreadis A, Snowden J, Craufurd D, Neary D, Owen F, Oostra BA, Hardy B, Goate A, Van Swieten J, Mann D, Lynch T, Heutink P (1998) Association of missense and 5'-splice-site mutations in tau with the inherited dementia FTDP-17. Nature 393:702–705

Kitada T, Asakawa S, Hattori N, Matsumine H, Yamamura Y, Minoshima S, Yokochi M, Mizuno Y, Shimizu N (1998) Mutations in the parkin gene cause autosomal recessive juvenile parkinsonism. Nature 392:605–608.

Klein C, Pramstaller PP, Kis B, Page CC, Kann M, Leung J, Woodward H, Castellan CC, Scherer M, Vieregge P, Breakefield XO, Kramer PL, Ozelius LJ (2000) Parkin deletions in a family with adult-onset, tremor-dominant parkinsonism: expanding the phenotype. Ann Neurol 48:65–71.

Klivenyi P, Ferrante RJ, Matthews RT, Bogdanov MB, Klein AM, Andreassen OA, Mueller G, Wermer M, Kaddurah-Daouk R, Beal MF (1999) Neuroprotective effects of creatine in a transgenic animal model of amyotrophic lateral sclerosis. Nature Med 5:347–350.

Levy-Lahad E, Wasco W, Poorkaj P, Romano DM, Oshima J, Pettingell WH, Yu CE, Jondro PD, Schmidt SD, Wang K, Crowley A, Fu YH, Guenette YS, Galas D, Nemens E, Wijsman EM, Bird TD, Schellenberg GD, Tanzi RE (1995) Candidate gene for the chromosome 1 familial Alzheimer's disease locus. Science 269:973–977.

Lewis J, Dickson DW, Lin WL, Chisholm L, Corral A, Jones G, Yen SH, Sahara N, Skipper L, Yager D, Eckman C, Hardy J, Hutton M, McGowan E (2001) Enhanced neurofibrillary degeneration in transgenic mice expressing mutant tau and APP. Science 293:1487–1491.

Lewis J, McGowan E, Rockwood J, Melrose H, Nacharaju P, Van Slegtenhorst M, Gwinn-Hardy K, Paul Murphy M, Baker M, Yu X, Duff K, Hardy J, Corral A, Lin WL, Yen SH, Dickson DW, Davies P, Hutton M (2000) Neurofibrillary tangles, amyotrophy and progressive motor disturbance in mice expressing mutant (P301L) tau protein. Nature Genet 25:402–405. Erratum in: Nature Genet (2000) 26:127

Lippa CF, Schmidt ML, Lee VM, Trojanowski JQ (2001) Alpha-synuclein in familial Alzheimer disease: epitope mapping parallels dementia with Lewy bodies and Parkinson disease. Arch Neurol 58:1817–1820.

Lippa CF, Smith TW, Saunders AM, Crook R, Pulaski-Salo D, Davies P, Hardy J, Roses AD, Dickson D (1995) Apolipoprotein E genotype and Lewy body disease. Neurology 45:97–103

Lloyd SE, Onwuazor ON, Beck JA, Mallinson G, Farrall M, Targonski P, Collinge J, Fisher EM (2001) Identification of multiple quantitative trait loci linked to prion disease incubation period in mice. Proc Natl Acad Sci USA 98:6279–6283

Lucking CB, Durr A, Bonifati V, Vaughan J, De Michele G, Gasser T, Harhangi BS, Meco G, Denefle P, Wood NW, Agid Y, Brice A (2000) Association between early-onset Parkinson's disease and mutations in the parkin gene. French Parkinson's Disease Genetics Study Group. N Engl J Med 342:1560–1567.

Masliah E, Rockenstein E, Veinbergs I, Mallory M, Hashimoto M, Takeda A, Sagara Y, Sisk A, Mucke L (2000) Dopaminergic loss and inclusion body formation in alpha-synuclein mice: implications for neurodegenerative disorders. Science 287:1265–129.

Masliah E, Rockenstein E, Veinbergs I, Sagara Y, Mallory M, Hashimoto M, Mucke L (2001) -amyloid peptides enhance alpha-synuclein accumulation and neuronal deficits in a transgenic mouse model linking Alzheimer's disease and Parkinson's disease. Proc Natl Acad Sci USA 98:12245–12250

McKeith IG, Galasko D, Kosaka K, Perry EK, Dickson DW, Hansen LA, Salmon DP, Lowe J, Mirra SS, Byrne EJ, Lennox G, Quinn NP, Edwardson JA, Ince PG, Bergeron C, Burns A, Miller BL, Lovestone S, Collerton D, Jansen EN, Ballard C, de Vos RA, Wilcock GK, Jellinger KA, Perry RH (1996) Consensus guidelines for the clinical and pathologic diagnosis of dementia with Lewy bodies (DLB): report of the consortium on DLB international workshop. Neurology 47:1113–1124

Mead S, Mahal SP, Beck J, Campbell T, Farrall M, Fisher E, Collinge J (2001) Sporadic--but not variant-Creutzfeldt-Jakob disease is associated with polymorphisms upstream of PRNP exon 1. Am J Human Genet 69:1225–1235

Medori R, Tritschler HJ, LeBlanc A, Villare F, Manetto V, Chen HY, Xue R, Leal S, Montagna P, Cortelli P, Tinuper P, Avoni P, Mochi M, Baruzzi A, Hauw JJ, Ott J, Lugaresi E, Autilio-Gambetti L, Gambetti P (1992) Fatal familial insomnia, a prion disease with a mutation at codon 178 of the prion protein gene. N Engl J Med 326:444–449

Muenter MD, Forno LS, Hornykiewicz O, Kish SJ, Maraganore DM, Caselli RJ, Okazaki H, Howard FM Jr, Snow BJ, Calne DB (1998) Hereditary form of parkinsonism-dementia. Ann Neurol 43:768–781.

Murayama O, Tomita T, Nihonmatsu N, Murayama M, Sun X, Honda T, Iwatsubo T, Takashima A (1999) Enhancement of amyloid beta 42 secretion by 28 different presenilin 1 mutations of familial Alzheimer's disease. Neurosci Lett 265:61–63

Myers A, Holmans P, Marshall H, Kwon J, Meyer D, Ramic D, Shears S, Booth J, DeVrieze FW, Crook R, Hamshere M, Abraham R, Tunstall N, Rice F, Carty S, Lillystone S, Kehoe P, Rudras-

ingham V, Jones L, Lovestone S, Perez-Tur J, Williams J, Owen MJ, Hardy J, Goate AM (2000) Susceptibility locus for Alzheimer's disease on chromosome 10. Science 290:2304–2305.
Olson MI, Shaw CM (1969) Presenile dementia and Alzheimer's disease in mongolism. Brain 92:147–156.
Ophoff RA, Terwindt GM, Vergouwe MN, van Eijk R, Oefner PJ, Hoffman SM, Lamerdin JE, Mohrenweiser HW, Bulman DE, Ferrari M, Haan J, Lindhout D, van Ommen GJ, Hofker MH, Ferrari MD, Frants RR (1996) Familial hemiplegic migraine and episodic ataxia type-2 are caused by mutations in the Ca2+ channel gene CACNL1A4. Cell 87:543–552.
Owen F, Poulter M, Lofthouse R, Collinge J, Crow TJ, Risby D, Baker HF, Ridley RM, Hsiao K, Prusiner SB (1989) Insertion in prion protein gene in familial Creutzfeldt-Jakob disease. Lancet 1:51–52.
Owen F, Poulter M, Shah T, Collinge J, Lofthouse R, Baker H, Ridley R, McVey J, Crow TJ (1990) An in-frame insertion in the prion protein gene in familial Creutzfeldt-Jakob disease. Brain Res Mol Brain Res 7:273–276.
Palmer MS, Dryden AJ, Hughes JT, Collinge J (1991) Homozygous prion protein genotype predisposes to sporadic Creutzfeldt-Jakob disease. Nature 352:340–342.
Pastor P, Roe CM, Villegas A, Bedoya G, Chakraverty S, Garcia G, Tirado V, Norton J, Rios S, Martinez M, Kosik KS, Lopera F, Goate AM (2003) Apolipoprotein Eepsilon4 modifies Alzheimer's disease onset in an E280A PS1 kindred. Ann Neurol 54:163–169.
Polymeropoulos MH, Lavedan C, Leroy E, Ide SE, Dehejia A, Dutra A, Pike B, Root H, Rubenstein J, Boyer R, Stenroos ES, Chandrasekharappa S, Athanassiadou A, Papapetropoulos T, Johnson WG, Lazzarini AM, Duvoisin RC, Di Iorio G, Golbe LI, Nussbaum RL (1997) Mutation in the alpha-synuclein gene identified in families with Parkinson's disease. Science 276:2045–2047.
Poorkaj P, Bird TD, Wijsman E, Nemens E, Garruto RM, Anderson L, Andreadis A, Wiederholt WC, Raskind M, Schellenberg GD (1998) Tau is a candidate gene for chromosome 17 frontotemporal dementia. Ann Neurol 43:815–825.
Prasher VP, Farrer MJ, Kessling AM, Fisher EM, West RJ, Barber PC, Butler AC (1998) Molecular mapping of Alzheimer-type dementia in Down's syndrome. Ann Neurol 43:380–383.
Prusiner SB (1998) Prions. Proc Natl Acad Sci USA 95:13363–13383.
Puls I, Jonnakuty C, LaMonte BH, Holzbaur EL, Tokito M, Mann E, Floeter MK, Bidus K, Drayna D, Oh SJ, Brown RH Jr, Ludlow CL, Fischbeck KH (2003) Mutant dynactin in motor neuron disease. Nature Genet 33:455–456.
Pulst SM (2003) Neurogenetics: single gene disorders. J Neurol Neurosurg Psychiat 74:1608–1614
Rippon GA, Crook R, Baker M, Halvorsen E, Chin S, Hutton M, Houlden H, Hardy J, Lynch T (2003) Presenilin 1 mutation in an african american family presenting with atypical Alzheimer dementia. Arch Neurol 60:884–888.
Rogaev EI, Sherrington R, Rogaeva EA, Levesque G, Ikeda M, Liang Y, Chi H, Lin C, Holman K, Tsuda T, Mar L, Sorbi S, Nacmias B, Placentini S, Amaducci L, Chumakov I, Cohen D, Lannfelt L, Fraser PE, Rommens JM, St George-Hyslop PH. (1995) Familial Alzheimer's disease in kindreds with missense mutations in a gene on chromosome 1 related to the Alzheimer's disease type 3 gene. Nature 376:775–778.
Rosen DR, Siddique T, Patterson D, Figlewicz DA, Sapp P, Hentati A, Donaldson D, Goto J, O'Regan JP, Deng HX, Rahmani Z, Krizus A, McKenna-Yasek D, Cayabyab A, Gaston SM, Berger R, Tanzi RE, Halperin JJ, Herzfeldt B, Van den Bergh R, Hung WY, Bird T, Deng G, Mulder DW, Smyth C, Laing NG, Soriano E, Pericak-Vance MA, Haines J, Rouleau GA, Gusella JS, Horvitz HR, Brown RH (1993) Mutations in Cu/Zn superoxide dismutase gene are associated with familial amyotrophic lateral sclerosis. Nature 362:59–62.
Scheuner D, Eckman C, Jensen M, Song X, Citron M, Suzuki N, Bird TD, Hardy J, Hutton M, Kukull W, Larson E, Levy-Lahad E, Viitanen M, Peskind E, Poorkaj P, Schellenberg G, Tanzi R, Wasco W, Lannfelt L, Selkoe D, Younkin S (1996) Secreted amyloid beta-protein similar to

that in the senile plaques of Alzheimer's disease is increased in vivo by the presenilin 1 and 2 and APP mutations linked to familial Alzheimer's disease. Nature Med 2:864–870.

Sherrington R, Rogaev EI, Liang Y, Rogaeva EA, Levesque G, Ikeda M, Chi H, Lin C, Li G, Holman K, Tsuda T, Mar L, Foncin JF, Bruni AC, Montesi MP, Sorbi S, Rainero I, Pinessi L, Nee L, Chumakov I, Pollen D, Brookes A, Sanseau P, Polinsky RJ, Wasco W, Da Silva HAR, Haines JL, Pericak-Vance MA, Tanzi RE, Roses AD, Fraser PE, Rommens JM, St George-Hyslop PH (1995) Cloning of a gene bearing missense mutations in early-onset familial Alzheimer's disease. Nature 375:754–760.

Shimura H, Hattori N, Kubo S, Mizuno Y, Asakawa S, Minoshima S, Shimizu N, Iwai K, Chiba T, Tanaka K, Suzuki T (2000) Familial Parkinson disease gene product, parkin, is a ubiquitin-protein ligase. Nature Genet 25:302–305.

Shimura H, Schlossmacher MG, Hattori N, Frosch MP, Trockenbacher A, Schneider R, Mizuno Y, Kosik KS, Selkoe DJ (2001) Ubiquitination of a new form of alpha-synuclein by parkin from human brain: implications for Parkinson's disease. Science 293:263–269.

Singleton AB, Farrer M, Johnson J, Singleton A, Hague S, Kachergus J, Hulihan M, Peuralinna T, Dutra A, Nussbaum R, Lincoln S, Crawley A, Hanson M, Maraganore D, Adler C, Cookson MR, Muenter M, Baptista M, Miller D, Blancato J, Hardy J, Gwinn-Hardy K (2003) α-Synuclein locus triplication causes Parkinson's disease.Science 302:841

Singleton A, Myers A, Hardy J (2004) The law of mass action applied to neurodegenerative disease: a hypothesis concerning the etiology and pathogenesis of complex diseases. Human Mol Genet 13 Spec No 1:R123–126.

Spillantini MG, Bird TD, Ghetti B (1998) Frontotemporal dementia and Parkinsonism linked to chromosome 17: a new group of tauopathies. Brain Pathol 8:387–402.

Subramony SH, Hernandez D, Adam A, Smith-Jefferson S, Hussey J, Gwinn-Hardy K, Lynch T, McDaniel O, Hardy J, Farrer M, Singleton A (2002) Ethnic differences in the expression of neurodegenerative disease: Machado-Joseph disease in Africans and Caucasians. Mov Disord 17:1068–1071

Tang-Wai D, Lewis P, Boeve B, Hutton M, Golde T, Baker M, Hardy J, Michels V, Ivnik R, Jack C, Petersen R (2002) Familial frontotemporal dementia associated with a novel presenilin-1 mutation. Dement Geriatr Cogn Disord 14:13–21.

Togo T, Sahara N, Yen SH, Cookson N, Ishizawa T, Hutton M, de Silva R, Lees A, Dickson DW (2002) Argyrophilic grain disease is a sporadic 4-repeat tauopathy. J Neuropathol Exp Neurol 61:547–556.

Valente EM, Abou-Sleiman PM, Caputo V, Muqit MM, Harvey K, Gispert S, Ali Z, Del Turco D, Bentivoglio AR, Healy DG, Albanese A, Nussbaum R, Gonzalez-Maldonado R, Deller T, Salvi S, Cortelli P, Gilks WP, Latchman DS, Harvey RJ, Dallapiccola B, Auburger G, Wood NW (2004) Hereditary early-onset Parkinson's disease caused by mutations in PINK1. Science 304:1158–1160.

Walker RH, Friedman J, Wiener J, Hobler R, Gwinn-Hardy K, Adam A, DeWolfe J, Gibbs R, Baker M, Farrer M, Hutton M, Hardy J (2002) A family with a tau P301L mutation presenting with parkinsonism. Parkinsonism Relat Disord 9:121–123.

Wavrant-De Vrieze F, Crook R, Holmans P, Kehoe P, Owen MJ, Williams J, Roehl K, Laliiri DK, Shears S, Booth J, Wu W, Goate A, Chartier-Harlin MC, Hardy J, Perez-Tur J (1999) Genetic variability at the amyloid-beta precursor protein locus may contribute to the risk of late-onset Alzheimer's disease. Neurosci Lett 269:67–70.

Wolfe MS, Xia W, Ostaszewski BL, Diehl TS, Kimberly WT, Selkoe DJ (1999) Two transmembrane aspartates in presenilin-1 required for presenilin endoproteolysis and gamma-secretase activity. Nature 398:513–517

Woodruff BK, Baba Y, Hutton ML, Wszolek ZK, Tsuboi Y, Kobayashi T, Ghetti B, Arima K, Yasuda M, Rascol O (2004) Haplotype-phenotype correlations in kindreds with the N279K mutation in the tau gene. Arch Neurol 61:1327

Yang Y, Hentati A, Deng HX, Dabbagh O, Sasaki T, Hirano M, Hung WY, Ouahchi K, Yan J, Azim AC, Cole N, Gascon G, Yagmour A, Ben-Hamida M, Pericak-Vance M, Hentati F, Siddique T (2001) The gene encoding alsin, a protein with three guanine-nucleotide exchange factor domains, is mutated in a form of recessive amyotrophic lateral sclerosis. Nature Genet 29:160–165.

Zarranz JJ, Alegre J, Gomez-Esteban JC, Lezcano E, Ros R, Ampuero I, Vidal L, Hoenicka J, Rodriguez O, Atares B, Llorens V, Gomez Tortosa E, del Ser T, Munoz DG, de Yebenes JG (2004) The new mutation, E46K, of alpha-synuclein causes Parkinson and Lewy body dementia. Ann Neurol 55:164–173.

Zhuchenko O, Bailey J, Bonnen P, Ashizawa T, Stockton DW, Amos C, Dobyns WB, Subramony SH, Zoghbi HY, Lee CC (1997) Autosomal dominant cerebellar ataxia (SCA6) associated with small polyglutamine expansions in the alpha(1A)-voltage-dependent calcium channel. Nature Genet 15:62–69

Racial and Ethnic Influences on the Expression of the Genotype in Neurodegenerative Diseases

Katrina Gwinn-Hardy[1]

Introduction

Although racial and ethnic variability and genetics are the subjects of considerable modern hype, the concepts underlying them are longstanding in clinical practice. Current and developing genetic techniques make it possible to study genomic variation between population groups and allow testing of the hypothesis that diseases have divergent clinical features between populations. There is growing evidence that race and ethnicity modulate disease via genetic background, although it is difficult to consider this separately from the influence of environment on the phenotype, since ethnicity may correlate with geographic and behavioral differences. Additionally, phenotypes differ within as well as across different racial and ethnic groups. Genetic, epidemiologic and pathologic investigations of neurodegenerative disease in different populations are required to further clarify the importance of these findings. Such investigations present challenges from a design standpoint and require consideration of political concerns. However, they are essential if we are to achieve appropriate diagnosis and treatment globally. The definition of the most appropriate treatment is that which addresses the underlying biology of disease. Once the diagnosis and treatment for neurodegenerative diseases are based on an understanding of pathogenesis, these differences in the phenotype will be of even greater clinical relevance.

Definitions of Race, Ethnicity

The terms "race," "ethnicity" and "ancestry" are often used interchangeably and overlap with the terms "culture" and "geographical isolates." It is likely that these terms will have some important distinctions from each other as the science of genetic epidemiology evolves. A recent *Nature Genetics* editorial defined race as one of the following:

- "a vague unscientific term for a group of genetically related people who share certain physical characteristics; a distinct ethnic group characterized by traits transmitted through offspring; each of the major divisions of humankind, having distinct physical characteristics; a group of individuals who are more or

[1] National Institute of Neurological Disorders and Stroke, National Institutes of Health, Bethesda, MD 20892 USA

Cummings et al.
Genotype – Protéotype – Phenotype
Relationships in Neurodegenerative Diseases

less isolated geographically or culturally, who share a common gene pool, and whose allele frequencies at some loci differ from those of other populations" (Editorial 2000).

That editorial further defined an ethnic group as:

- "a population of individuals organized around an assumption of common cultural origin or those individuals with a common national or cultural tradition: a social group or category of the population that, in a larger society, is set apart and bound together by common ties of race, language, nationality or culture" (Editorial 2000).

It is notable that the definition of race includes genetic features prominently. Both race and ethnicity have relevance for genetic studies, especially genotype-phenotype studies. However, how these terms are best applied in biomedical research remains to be further refined.

Complex Disease

Racial and Ethnic Differences in Classical Medical Teaching

That disease may appear differently in different racial groups is not a new concept to medicinal practice. When a clinician considers treatment based on gender or family history, he/she is, broadly speaking, using genetic principles, since gender and family history are determined by genotype. Individuals with a family history of heart disease or colon cancer, for example, will undergo more aggressive screening than those who do not. Asking patients to list their allergies is also a form of rudimentary genetic inquiry, since many allergies "run in families" and are likely due to genetic variations. This tendency is particularly well known in the field of anesthesiology, where a family history of a "bad reaction" to anesthesia is considered a red flag prior to surgery. It is widely accepted among clinicians that some racial and ethnic groups differ significantly in terms of incidence and mortality rates from different types of disease. For example, black Americans have a higher risk of esophageal cancer, multiple myeloma, and stomach cancer, whereas white Americans have a higher incidence of melanoma, leukemia, and some lymphomas (Perera 1997). Note that, because exposures may also differ between these groups, studies designed to determine the effect of ethnicity on these phenotypes are methodologically complicated.

There are clear cases of Mendelian mutations leading to disease in primarily a single racial group. Examples include hereditary hemochromatosis due to Cys282Tyr mutation homozygosity, which is rare among Indians and Chinese but occurs in 7.5 % of Swedes (Beckman et al. 1997). Sickle cell anemia is prevalent among African origin people; no one would fault a clinician for ordering a lab test for sickle cell anemia only on black Americans with stroke. These examples are some among many, which not only underscore the importance of clinical and genetic differences across racial and ethnic groups but also their broad clinical utility. A particularly interesting example is that of the BRCA mutation, which increases the risk of early-onset breast cancer in females and is more common in Ashkenazi Jews; a significant minority of Jewish women surveyed have grave concerns regarding testing based on ethnicity (Lehmann et al. 2002). The controversies surrounding genetic testing of Jewish women and prophylactic mastectomy in those who have mutations will inform decision-making processes in other disease areas as our knowledge increases.

Ethnicity and race: political or scientific?

How terms are defined depends on their intended use and scope. The *Nature Genetics* definition (above; Editorial 2000) will guide science (see also Rosenberg et al. 2002; Balter and Gibbons 2003). Other definitions have been developed for the formulation of public policies and other governmental and societal functions (Schwartz 2001). What is the best means of racial and/or ethnic identification? Recently, self-identification has become the standard. However, even self-identification of race can be problematic, particularly in a country such as the United State of America where admixture is common. The Office of Management and Budget of the US allows multiple responses to a question on racial identification in the US Census. According to a *New York Times* article on the subject, about seven million people identified themselves as members of more than one racial group in the 2000 census (Schmitt 2001). Furthermore, about 800,000 people in that same census apparently defined themselves as being both white and black (Schmitt 2001). This degree of multiracial identification underscores the heterogeneity of the US population and reveals how difficult it might be to ask a scientific question regarding ethnic or racial identity in the existing culture of US society.

Only relatively recently has there been adequate recognition of the fact that, in the past, certain subgroups of people were not adequately represented in clinical research, including in clinical trials. To address this significant concern, the US National Institutes of Health (NIH) has developed a policy on the inclusion of women and minorities as subjects in clinical research, http://grants.nih.gov/grants/funding/women_min/guidelines_amended_10_2001.htm, which states that all NIH-funded clinical research must be carried out in a manner sufficient to elicit information about individuals of both sexes/genders and diverse racial and ethnic groups. Additionally, particularly in NIH-defined Phase III clinical trials, trials must include women and minorities in order to examine differential effects on such groups. Since a primary aim of research is to provide scientific evidence leading to a change in health policy or standard of care, it is imperative to determine whether the intervention or therapy being studied affects women, men, or members of minority groups and their subpopulations differently. It remains unclear, however, how subset analysis should be considered (Brennan 1999; Iannuzzi et al. 2002; Wacholder et al. 2000). A principal tenet of clinical trials design is that the intervention should not be expected to operate differently among the various subsets of participants (Friedman et al. 1998). This tenet raises a methodological issue: additional subgroup analyses require larger sample sizes and run the risk of erroneous conclusions due to multiple testing. It is likely that these concerns will become increasingly important as the realm of clinical trials expands towards pharmacogenetics, which will require the evolution of both clinical trials methodology and public policy.

Diagnostic Criteria for Neurodegenerative Diseases

How Is a Diagnosis of a Neurodegenerative Disorder Reached?

The brain is not generally examined directly during life; exceptions include biopsy in the case of tumors, some infections, and suspected prion disease. Classically, diagnostic criteria were developed once the clinical entity was defined and those entities were correlated with the neuropathological findings felt to be pathognomonic. Therefore, diagnoses of neurodegenerative disease are made post-mortem, when the brain can be examined and findings can be correlated with a history of similar symptom constellations during life. These defining studies were largely done in Caucasians and in the Japanese population, to a lesser degree. With genetic testing now available, there are exceptions where diagnoses can be made during life, and even before the onset of symptoms. When these molecular diagnoses do not fit with what is expected clinically or pathologically, dogma is challenged, which is disconcerting to the academic community.

An important example showing the shift in how neurodegenerative diagnosis is made is Pick's disease. Classically, Pick's disease is defined as a dementing illness during lifetime that has the pathologicial features of frontotemporal lobar atrophy and argyrophilic inclusion bodies within the cytoplasm (Pick's bodies; Bradley et al. 1991). Pick's disease, then, is classically a clinicopathological entity. As genetic information on this disorder expanded, the disease was enveloped in the broader term, frontotemporal dementia, linked to chromosome 17 (FTDP-17). This renaming was precipitated by an international consensus conference that brought researchers together who had studied 13 apparently unrelated kindreds whose autosomal dominant neurodegeneration syndrome showed linkage to chromosome 17 (Foster et al. 1997). Ultimately, mutations in the tau gene revealed a single genetic cause. (Hutton et al. 1998; Spillantini et al. 1998). This shift from neuropathological diagnosis to genetic or underlying cellular process, analysis as a means of diagnosis has allowed us to rethink our classic categorizations. Part of this rethinking is the idea that a given genotype leads to a rigidly defined phenotype.

It is difficult to determine whether disease phenotypes, used for gene discovery, diagnosis, and treatment, are similar across racial and ethnic groups because of the paucity of clinicopathological studies in non-Caucasians. This lack of information has led to a fundamental, although possibly flawed, assumption in epidemiological cross-cultural comparisons of disease: that clinically defined disorders are similar in cause and outcome across racial and ethnic groups (Hardy et al. 2003).

Influences of Race and Ethnicity on the Phenotype

Monogenic Neurodegenerative Diseases as an Example

Taking monogenic neurodegenerative diseases as an example, one notes that, as reports of disease in various racial and ethnic groups increase, there are increasing numbers of disparate phenotypes between these groups.

The spinocerebellar ataxia (SCA) 2 gene was discovered on chromosome 12q in 1996 (Pulst et al. 1996; Sanpei et al. 1996). The classic phenotype is that of ataxia with other features, including ophthalmoplegia and peripheral neuropathy; this phenotype is known to be variable not only across racial and ethnic groups but also across, and even within, families (Gwinn-Hardy et al. 2000). SCA2 mutations usually present with a pure ataxia syndrome in Caucasians (Schols et al. 1997), but in some Chinese families a phenotype occurs that is almost indistinguishable from Parkinson's disease (PD) clinically and on PET neuroimaging (Table 1; Gwinn-Hardy et al. 2000; Shan et al. 2001; Lu et al. 2002). As noted in Table 1, the phenotype within a given family, the Taipei Kindred, ranges from ataxia to mixed ataxia-parkinsonism to "pure" parkinsonism (Gwinn-Hardy et al. 2000). As repeat sizes increase, the phenotype more closely resembles ataxia. In commenting on these findings, Kock et al. (2002) stated that, in a study of 270 unrelated, largely Caucasian and a few east Indian PD patients, they found no expanded SCA2 alleles, confirming that genetic contributions to a given clinical diagnostic entity differs, across racial and ethnic populations. Subsequently, Lu et al. (2002) revealed that, in their series of ethnic Chinese in Taiwan, SCA2 is responsible for almost 10% of familial parkinsonism cases. Pathological studies of SCA2 (done in Caucasians) show degeneration of the olivopontocerebellar regions as well as severe neuronal loss in the substantia nigra, striatum, pallidum, and even neocortex (Estrada et al. 1999). Thus, one would not be surprised to find parkinsonism, and even dementia, as a phenotypic manifestation in addition to that of ataxia even in Caucasians. Such is indeed the case (Furtado et al. 2002). Clearly, this difference is quantitative rather than qualitative, since nigral loss probably occurs in all SCA2 cases, but in Caucasians this nigral loss may not typically be enough to be clinically relevant. Related observations have been made in SCA3 mutation carriers (Rosenberg and Fowler 1981). As in SCA2, in SCA3 (Machado Joseph disease), affected members of Caucasian families typically present with ataxia, although mixed ataxia-parkinsonism is also seen (Tuite et al. 1995): in contrast, in people of African origin with SCA3, ataxia overlapped by parkinsonism in a given patient is common (Table 2; Gwinn-Hardy et al. 2001; Subramony et al. 2002). Chinese kindreds in whom affected members have typical PD have also been found with SCA3 expansions (Chung et al. 2003).

It is notable that the PD phenotype for both SCA2 and 3 may be more common in those with intermediate-range repeat numbers; this narrow range requires precise determination of repeat size, which is not done by all diagnostic laboratories (Hussey et al. 2002). Therefore, when ruling out SCA2 and SCA3 as a cause of PD, it is important that methods that accurately assess the repeats in this intermediate (a.k.a. borderline) range be used.

Table 1. Clinical features in SCA2 Chinese subjects

Taipei Kindred SCA2				
Subject #	Ataxia features	Parkinsonism	Dopamine esponsive	Diagnosis
2001	y	y	u	Ataxia
3000	y	y	u	u
3009	y	y	u	PD
4015	y	y	y	PD
3013	n	y	y	PD
4017	n	y	y	PD
4002	n	y	n	PSP

SCA2, spinocerebellar ataxia type 2; SCA3, spinocerebellar ataxia type 3; PD, Parkinson's disease; PSP, progressive aupranuclear palsy; u, unknown. Diagnoses were clinical diagnoses by the subjects' neurologist. Those with ataxia are shaded in red, mixed features in tangerine, and Parkinsonism, including PD and PSP, in lemon (see also Gwinn-Hardy et al. 2001, 2002).

Other examples include in Dentatorubropallidoluysian Atrophy (DRPLA) and Alzheimer's disease (AD). The DRPLA phenotypes in Japanese and European populations and Haw River syndrome in African Americans were disparate enough to hide the fact these diseases were caused by the same gene mutation (Burke et al. 1994; Nagafuchi et al. 1994). Presenilin mutations, known to cause typical, albeit early onset AD, have been described widely in Caucasian and Japanese patients, usually causing predominant memory loss (Warrington et al. 2001), but in the single reported example of a presenilin mutation occurring in an African kindred, personality changes and disinhibition were early features (Rippon et al. 2003).

Note, however, that none of the above discussed studies are population based. To truly determine if these trends are reflective of an overall racial difference, genetic epidemiology studies are needed. If these findings are, indeed, verified across populations, there would be a significant impact on our understanding of the incidence and prevalence of neurodegenerative disease categories. Although the burden of disease may be similar between racial and ethnic groups, the clinical diagnoses may differ. Studies of diseases such as AD and PD, diagnosed based on consensus criteria designed in Caucasians, may give lower rates of disease in other populations (because true cases will be "missed" that are phenotypically outside the traditional syndrome). These criteria have been fine-tuned to exclude all "similar" diseases, which may actually be the clinical manifestations of that same disease in another population. A recent review (McInerney-Leo et al. 2004) noted that many articles on PD state that this disease is less common in African

Table 2. Clinical features in SCA3 African American/Caribbean and Caucasian families (multiple kindreds)

SCA3 African American, Caribbean

Subject	Ataxia features	Parkinsonism	Dopamine responsive	Diagnosis
5	y	n	u	Ataxia
7	y	y	u	Ataxia
20	y	y	u	Ataxia
4	y	y	u	Ataxia
22	y	y	u	PD
9719	y	y	u	PD
19	y	y	y	PD
9717	n	y	u	u
21	y	y	y	PD
9718	y	y	y	PD
2001	n	y	y	PD
3002	n	y	y	PD
3012	n	y	y	PD
018	n	n	u	dystonia

SCA3 Caucasian

Subject # features	Ataxia	Parkinsonism responsive	Dopamine	Diagnosis
695	y	n	u	Ataxia
15	y	n	u	Ataxia
2	y	n	u	Ataxia
1	y	n	u	Ataxia
23	y	y	u	Ataxia
700	y	y	u	Ataxia

SCA2, spinocerebellar ataxia type 2; SCA3, spinocerebellar ataxia type 3; PD, Parkinson's Disease; PSP, progressive supranuclear palsy; u, unknown. Diagnoses were clinical diagnoses by the subjects' neurologist. Those with ataxia are shaded in red, mixed features in tangerine, and Parkinsonism, including PD and PSP, in lemon (see also Gwinn-Hardy et al. 2001; Subramony ct al. 2002).

origin people than in Caucasians. Additional study is needed, but, one reason this might appear to be the case could be that alpha-synucleinopathy, which is probably a major cause of typical PD in Caucasians, may present with different diagnostic features in non-Caucasians. There is a call for appropriate neuropathological and genetic epidemiological studies across all racial groups to answer some basic questions regarding genotype-phenotype.

Therapeutic Implications

Of great importance are the therapeutic implications of racial and ethnic differences in phenotype. Patients with similar clinical phenotypes might need different treatments if that phenotype has an entirely different molecular cause. Once therapies are based on an understanding of pathogenesis, these differences will need to be recognized.

Some argue against the use of racial/ethnic categories in genetic and other clinical research studies because of the risk of furthering divisiveness and allowing perpetuation of harmful stereotypes that might arise from classifying people by race/ethnicity (Bogue and Edwards 1971; Terris 1973; Azuonye 1996; Fullilove 1998). Others have argued that documenting health disparities justifies continued racial and ethnic categorization of clinical study participants (LaVeist 1996; Williams 1997; Kreiger et al. 1999; Srinivasan and Guillermo 2000; Flores 2001). An additional argument in favor of continued classification is that it may aid the study of pharmacogenetics and genetic epidemiology. A "race-neutral" approach to genetic research would not reveal the influence of race on the phenotype. Additionally, the above data suggest that such an approach would increase the risk of misclassification of some neurodegenerative disease entities. Determining if Caucasians, African Americans, Hispanics, Native Americans, Pacific Islanders or Asians respond similarly to a given treatment requires that these groups be considered individually (Risch et al. 2002). Careful consideration must be given in clinical studies to the way in which race and ethnicity are conceptualized, the choice and definition of categories, and the way in which individuals are assigned to categories. It is likely that the definitions given by *Nature Genetics*, which are extremely valuable, will become even more refined as our sensitivity grows.

Summary and Recommendations for Directions

The above discussion is important for several reasons. First, it is important genetically; it suggests that there are modifier genes that subtly alter disease phenotypes. As we understand the pathogeneses of these diseases more clearly, we should expect some of the genes in the pathogenic pathways to have different allele frequencies in different racial groups.

Second, it is important diagnostically and therapeutically. The classical pathway of disease definition had been to clarify a constellation of symptoms, as James Parkinson did at the turn of the century. Subsequently, pathological correlates were sought. Thereafter, gene discovery took place, based on a single "diagnosis," i.e., a constellation of symptoms and signs to allow categorization during life and predict natural history and treatment response. Future treatments for neurodegenerative diseases are likely to be based on molecular pathogenesis, not merely palliative measures, as they are currently. This approach will allow treatment to stem from the underlying biology of disease and, hence, to tackle the disease at its root cause rather than in a series of what might be common final pathways for a variety of biological processes, all leading to similar or the same phenotype. Patients with similar clinical phenotypes might need very different treatments, depending on genetic cause or risk factors. For example, based on the above-discussed data, treatment decisions in ataxia will be relevant to a subset of PD patients, and the size of this subset will differ in differing populations. This is likely to be true for a variety of disorders. Inclusion/exclusion criteria for clinical studies may need to consider molecular characterizations as well, in order to refine diagnostic categories and allow more sophisticated treatment response analyses.

We need additional genetic, epidemiologic and pathologic investigations of neurodegenerative disease in different populations. In these investigations, it will be inappropriate to naïvely and uniformly apply consensus criteria derived from the analysis of Caucasian populations: rather, criteria will have to be considered in many ethnic and racial groups. Such investigations, despite their challenges from a design standpoint, are essential if we are to achieve appropriate diagnosis and treatment for non-Caucasian and Caucasian populations alike.

Several studies have looked at a given disease and, within that diagnosis, determined differences in genotype for differing manifestations. It is time to tackle this from the opposite angle: by using genetic factors to define the underlying biology, we can broaden our understanding, both clinical and, ultimately, pathological, of the possible phenotypes that might be manifest as a result of a given genetic variant.

These differences in phenotype that are influenced by racial and ethnic background are not silos: they do not stand in isolation from other diagnostic entities. Though admixture could certainly explain the overlap in phenotypes in Caribbean black families and Caucasian families (Gwinn-Hardy et al. 2001; Subramony et al. 2002), such admixture cannot explain the overlap in the Mendelian families in Taiwan (ethnic Chinese; Tables 1, 2). In these cases, the phenotypic range is mostly the same as that of Caucasians: the differences lie at the edges of the phenotype. However, these edges are extremely important from a public health standpoint, as we may be making diagnoses that we suspect reflect one underlying biological process but actually reflect another.

A major discussion has arisen recently regarding optimal strategies for racial categorization, especially in the United States, for the purpose of biomedical research. (Risch et al. 2002) It is important to know whether particular individuals within the population are more susceptible to particular diseases or are most likely to benefit from certain therapeutic interventions. The data reveal, paradoxically, that both are right: Racial and ethnic divisions are meaningless, since diseases can and often do look the same across racial and ethnic groups; conversely, racial and ethnic variability can influence the phenotype in ways that can have dramatic therapeutic implications. As inscribed on the frieze on the facade of the

National Academy of Sciences building, in Washington DC, USA, the advice from Aristotle's "Metaphysics" is fitting indeed. In English, it reads:

- "The investigation of truth is in one way hard and in another way easy. An indication of this is found in the fact that no one is able to attain the truth entirely, while on the other hand no one fails entirely, but everyone says something true about the nature of things, and by the union of all a considerable amount is amassed." (Kelly 2004)

A single gene or genetic risk factor can no longer simplistically be considered associated with a single phenotype: the phenotype is likely to vary based on many factors, including ethnicity and race. Genotype-phenotype studies are neither linear nor even bi-directional but spiral, informing each other and allowing ongoing refinement of our understanding of each.

References

Azuonye IO (1996) Guidelines will encourage the thinking that underpins racism in medicine. Brit Med J 313:426.

Balter M, Gibbons A (2000) Paleoanthropology. A glimpse of humans' first journey out of Africa. Science 288: 948–950.

Beckman LE, Saha N, Spitsyn V, Van Landeghem G, Beckman L (1997) Ethnic differences in the HFE codon 282 (Cys/Tyr) polymorphism. Human Hered 47:263–267.

Bogue G, Edwards GF (1971) How to get along without race in demographic analysis. Soc Biol 18:387–396.

Bradley WG, Daroff RB Fenichel GM, Marsden CD (1991) Neurology in clinical practice. The neurological disorders. Volume II. Butterworth-Heinemann, Boston, p. 1417.

Brennan P (1999) Design and analysis issues in case-control studies addressing genetic susceptibility. IARC Sci Publ 148:123–132.

Burke JR, Wingfield MS, Lewis KE, Roses AD, Lee JE, Hulette C, Pericak-Vance MA, Vance JM (1994) The Haw River syndrome: dentatorubropallidoluysian atrophy (DRPLA) in an African-American family. Nature Genet 7:521–524.

Chung MY, Lu YC, Cheng NC, Soong BW (2003) A novel autosomal dominant spinocerebellar ataxia (SCA22) linked to chromosome 1p21-q23. Brain 126:1293–1299

Editorial (2000) Census, race and science. Nature Genet 24:97–98.

Estrada R, Galarraga J, Orozco G, Nodarse A, Auburger G (1999) Spinocerebellar ataxia 2 (SCA2): morphometric analyses in 11 autopsies. Acta Neuropathol (Berl) 97:306–310.

Flores G, Fuentes-Afflick E, Carter-Pokras O, Claudio L, Lamberty G, Lara M, Pachter L, Ramos Gomez F, Mendoza F, Valdez RB, Zambrana RE, Greenberg R, Weitzman M. (2001) Why ethnicity and race are so important in child health services research today. Arch Pediatr Adolesc Med 155:1178–1179.

Foster NL, Wilhelmsen K, Sima AA, Jones MZ, D'Amato CJ, Gilman S (1997) Frontotemporal dementia and parkinsonism linked to chromosome 17: a consensus conference. Conference Participants. Ann Neurol 41:706–715.

Fullilove MT (1998) Comment: abandoning "race" as a variable in public health research: an idea whose time has come. Am J Publ Health 88:1297–1298.

Friedman LM, Furberg CD, DeMets, DL, eds (1998) Fundamentals of Clinical Trials. 3rd edition. Springer, New York.

Furtado S, Farrer M, Tsuboi Y, Klimek ML, de las Fuente-Fernandez R, Hussey J, Lockhardt P, Calne DB, Suchowersky O, Stoessl AJ, Wszolek Z (2002). SCA-2 presenting as parkinsonism in an Alberta family: clinical, genetic and PET findings. Neurology 59:1625–1627.

Gwinn-Hardy K, Chen, JY, Liu H.-C, Liu TY, Boss M, Seltzer W, Adam A, Singleton A, Koroshetz W, Waters C. Hardy J. Farrer M (2000) Spinocerebellar ataxia type 2 with parkinsonism in ethnic Chinese. Neurology 55: 800–805.
Gwinn-Hardy K, Singleton A, O'Suilleabhain P, Boss M, Nicholl D, Adam A, Hussey J, Critchley P, Hardy J, Farrer M (2001) Spinocerebellar ataxia type 3 phenotypically resembling Parkinson disease in a black family. Arch Neurol; 58:296–299.
Hardy J, Singleton A, Gwinn-Hardy K (2003) Ethnic differences and disease phenotypes. Science 300:739–740.
Hussey J, Lockhart PJ, Seltzer W, Wszolek ZK, Payami H, Hanson M, Gwinn-Hardy K, Farrer M (2002) Accurate determination of ataxin-2 polyglutamine expansion in patients with intermediate-range repeats. Genet Test 6:217–220.
Hutton M, Lendon CL, Rizzu P, Baker M, Froelich S, Houlden H, Pickering-Brown S, Chakraverty S, Isaacs A, Grover A, Hackett J, Adamson J, Lincoln S, Dickson D, Davies P, Petersen RC, Stevens M, de Graaff E, Wauters E, van Baren J, Hillebrand M, Joosse M, Kwon JM, Nowotny P, Che LK, Norton J, Morris JC., Reed LA, Trojanowski J, Basun H, Lannfelt L, Neystat M, Fahn S, Dark F, Tannenberg T, Dodd PR, Hayward N, Kwok JBJ, Schofield PR, Andreadis A, Snowden J, Craufurd D, Neary D, Owen F, Oostra BA, Hardy J, Goate A, van Swieten J, Mann D, Lynch T, Heutink P. (1998) Coding and splice donor site mutations in tau cause autosomal dominant dementia (FTDP-17). Nature 393:702–705.
Iannuzzi MC, Maliarik M, Rybicki B (2002) Genetic polymorphisms in lung disease: bandwagon or breakthrough? Resp Res 3:15:1186.
Kelly J (2004) Answer man: the word on science. Washington Post, Monday, September 6, C11.
Kock N, Muller B, Vieregge P, Pramstaller PP, Marder K, Abbruzzese G, Martinelli P, Lang AE, Jacobs H, Hagenah J, Harris J, Meija-Santana H, Fahn S, Hedrich K, Kann M, Gehlken U, Culjkovic B, Schwinger E, Wszolek ZK, Zuhlke C, Klein C (2002) Role of SCA2 mutations in early- and late-onset dopa-responsive parkinsonism. Ann Neurol 52: 257–258.
Krieger N, Williams D, Zierler S (1999) "Whiting out" white privilege will not advance the study of how racism harms health. Am J Publ Health 89:782–783.
LaVeist TA (1996) Why we should continue to study race – but do a better job: an essay on race, racism and health. Ethn Dis 6:21–29.
Lehmann LS, Weeks JC, Klar N, Garber JE (2002) A population-based study of Ashkenazi Jewish women's attitudes toward genetic discrimination and BRCA1/2 testing. Genet Med 4:346–352.
Lu CS, Wu Chou YH, Yen TC, Tsai CH, Chen RS, Chang HC (2002) Dopa-responsive parkinsonism phenotype of spinocerebellar ataxia type 2. Mov Disord 17:1046–1051.
McInerney-Leo A, Gwinn-Hardy K, Nussbaum RL (2004) Prevalence of Parkinson's disease in populations of African ancestry: a review. J Natl Med Assoc 96:974–979.
Nagafuchi S, Yanagisawa H, Sato K, Shirayama T, Ohsaki E, Bundo M, Takeda T, Tadokoro K, Kondo I, Murayama N, Tanaka Y, Kikushima H, Umino K, Kurosawa H, Furukawa T, Nihei K, Inoue T, Sano A, Komure O, Takahashi M, Yoshizawa T, Kanazawa I, Yamada M (1994) Dentatorubral and pallidoluysian atrophy expansion of an unstable CAG trinucleotide on chromosome 12p. Nature Genet 6:14–18.
Perera, FP (1997) Environment and cancer: who are susceptible? Science 278: 1068–1073.
Pulst SM, Nechiporuk A, Nechiporuk T, Gispert S, Chen X-N, Lopes-Cendes I, Pearlman S, Starkman S, Orozco-Diaz G, Lunkes A, DeJong P, Rouleau GA, Auburger G, Korenberg JR, Figueroa C, Sahba S (1996) Moderate expansion of a normally biallelic trinucleotide repeat in spinocerebellar ataxia type 2. Nature Genet 14: 269–276.
Rippon G, Baker M, Crook R, Hutton M, Hardy J, Lynch T (2003) An African American family with a presenilin mutation. Arch Neurol 60:884–888.
Risch N, Burchard E, Ziv E, Tang H (2002) Categorization of humans in biomedical research: genes, race and disease. Genome Biol 1;3: comment 2007. Epub Jul 01, 2002
Rosenberg, RN, Fowler, HL (1981) Autosomal dominant motor system disease of the Portuguese: a review. Neurology 31:1124–1126.

Rosenberg NA, Pritchard JK, Weber JL, Cann HM, Kidd KK, Zhivotovsky LA, Feldman MW (2002) Genetic structure of human populations. Science 298:2381–2385.
Sanpei K, Takano H, Igarashi S, Sato T, Oyake M, Sasaki H, Wakisaka A, Tashiro K, Ishida Y, Ikeuchi T, Koide R, Saito M, Sato A, Tanaka T, Hanyu S, Takiyama Y, Nishizawa M, Shimizu N, Nomura Y, Segawa M, Iwabuchi K, Eguchi I, Tanaka H, Takahashi H, Tsuji S (1996) Identification of the spinocerebellar ataxia type 2 gene using a direct identification of repeat expansion and cloning technique, DIRECT. Nature Genet 14:277–284.
Schmitt E (2001) For 7 million people in census, one race category isn't enough. New York Times. March 13, 2001:1.
Schols L, Gispert S, Vorgerd M, Menezes Vieira-Saecker AM, Blanke P, Auburger G, Amoiridis G, Meves S, Epplen JT, Przuntek H, Pulst SM, Riess O (1997) Spinocerebellar ataxia type 2. Genotype and phenotype in German kindreds. Arch Neurol 54:1073–1080.
Schwartz RS (2001) Racial profiling in medical research N Engl J Med. 18:1392–1393.
Shan D-E, Soong B-W, Sun C-M, Lee S-J, Liao K-K, Liu R-S (2001) Spinocerebellar ataxia type 2 presenting as familial levodopa-responsive parkinsonism. Ann Neurol 50:812–815.
Soong B, Cheng C, Liu R, Shan D (1997) Machado-Jospeh disease: clinical, molecular, and metabolic characterization in Chinese kindreds. Ann Neurol 41: 446–452.
Spillantini MG, Bird TD, Ghetti B (1998) Frontotemporal dementia and Parkinsonism linked to chromosome 17: a new group of tauopathies. Brain Pathol 8:387–402
Srinivasan S, Guillermo T (2000) Toward improved health: disaggregating Asian American and Native Hawaiian/Pacific Islander data. Am J Publ Health 90:1731–1734.
Subramony SH, Hernandez D, Adam A, Smith-Jefferson S, Hussey J, Gwinn-Hardy K, Lynch T, McDaniel O, Hardy J, Farrer M, Singleton A (2002) Ethnic differences in the expression of neurodegenerative disease: Machado-Joseph disease in Africans and Caucasians. Mov Disord 17:1068–1071.
Terris M (1973) Desegregating health statistics. Am J Publ Health 63:477–480.
Tuite PJ, Rogaeva EA, St George-Hyslop PH, Lang AE (1995) Dopa-responsive parkinsonism phenotype of Machado-Joseph disease: confirmation of 14q CAG expansion. Ann Neurol 38:684–687.
Wacholder S, Rothman N, Caporaso N (2000) Population stratification in epidemiologic studies of common genetic variants and cancer: quantification of bias. J Natl Cancer Inst. 92:1151–1158.
Warrington EK, Agnew SK, Kennedy AM, Rossor MN (2001) Neuropsychological profiles of familial Alzheimer's disease associated with mutations in the presenilin 1 and amyloid precursor protein genes. J Neurol 248:45–50.
Williams DR (1997) Race and health: basic questions, emerging directions. Ann Epidemiol 7:322–333.

Causes and Consequences of Oxidative Stress in Neurodegenerative Diseases

Domenico Praticò[1]

Summary

As people worldwide live to an older age, dementia, whose best-known risk factor is aging, has become a serious growing public health problem. While Alzheimer's disease (AD) is the most frequent cause of dementia, other progressive neurodegenerative disorders can be responsible for it. For these reasons, they are grouped as non-AD dementias. There is evidence that reactive oxygen species-mediated reactions, particularly of neuronal lipids, are present in brains affected by both types of dementing disorders. Traditional views have claimed that oxidative-mediated brain injury in these diseases is merely the result of the neurodegenerative processes. While numerous investigations have shown that oxidative stress is increased in AD, conflicting results exist for the heterogeneous group of non-AD dementias. The availability of specific and sensitive markers to monitor in vivo oxidative stress, in combination with studies performed in living patients, are helping us to elucidate these issues. This paper summarizes some of the most recent research on the relevance of oxidative stress and lipid peroxidation in AD and non-AD dementias. The evidence accumulated so far clearly indicates that oxidative stress is an early and specific aspect of AD pathogenesis but not of the pathogenesis of other dementias. This new concept implies that this phenomenon is not a general and common pathway of the neurodegenerative process, but it may play a more specific and important role in AD than in non-AD dementias.

Introduction

As people worldwide live to an older age, dementia, whose best-known risk factor is aging, has become a serious growing public health problem. Alzheimer's disease (AD) is the most frequent form of neurodegenerative disease associated with dementia in the elderly (Clark 2000). Aging, or senescence, the strongest risk factor for developing AD, is a typical feature of the post-reproductive phase of life. It manifests in all multinuclear organisms and is characterized by a progressive reduction in the efficacy of a number of physiological processes. This decline translates into a reduced capacity to maintain homeostatic control of important

[1] Center For Experimental Therapeutics, Department of Pharmacology, University of Pennsylvania, School of Medicine, Philadelphia, PA, 19104 John Morgan Building, room 124

Cummings et al.
Genotype – Protéotype – Phenotype
Relationships in Neurodegenerative Diseases

functions and finally results in death of the organism (Praticò 2002). Almost 50 years ago, Deham Harman (1956) proposed that the aging process is the result of a free radical-mediated damage to tissues or organs. At the core of this theory is the hypothesis that a progressive accumulation of oxidative damage is the fundamental cause underlying senescence-associated alterations. Following this theory, it has been proposed that oxidative stress may also be of functional importance in the pathogenesis of several neurodegenerative disorders (Praticò and Delanty 2000). In a situation of oxidative imbalance, an excessive production of reactive oxygen species (ROS) in the brain would lead to oxidative stress and possibly, as a final result, to neuron degeneration. The brain is highly sensitive to oxidative stress since it is very rich in easily peroxidizable fatty acids, has a high request for oxygen, and has a relative paucity of antioxidant systems. Furthermore, it has a high content of transition metals, such as iron, and high ascorbate levels, which together act as potent pro-oxidants (Floyd 1999).

Depending on the substrate attacked by ROS, oxidative stress will manifest as protein oxidation, DNA oxidation or lipid peroxidation. All of these signature markers of oxidative damage have been described in the AD brain, and a role for it has been widely discussed in the pathogenesis of AD (Smith et al. 2000; Christen 2000). In general, oxidative damage in the central nervous system predominantly manifests as lipid peroxidation, because the brain has a very high lipid content and, in particular, high levels of polyunsaturated fatty acids that are easily susceptible to ROS attack (Praticò 2002). However, oxidative stress and subsequent lipid peroxidation could also be secondary to neuronal degeneration per se. Despite this evidence, efforts to elucidate the role of lipid peroxidation in vivo have been hampered by the limited availability of markers, which have been of marginal value for research or diagnosis owing to their chemical instability or their lack of sensitivity or specificity (Halliwell and Gutteridge 1999). F_2-isoprostanes are a new class of lipids produced by free radical-catalyzed peroxidation of free or esterified fatty acids present in cell membranes, from which they can be cleaved, can circulate in blood and can be excreted in urine (Praticò et al. 2001). Differently from other lipid peroxidation products, they are chemically stable and present in detectable amounts in all normal biological fluids and tissues (Praticò et al. 2001). For these reasons, they provide a reliable index for assessing the oxidant component of several diseases in vivo, and growing amounts of data have accumulated clearly showing that they are specific and sensitive markers of lipid peroxidation.

Oxidative stress in AD

A considerable number of studies have been published on lipid peroxidation and AD. Most have been performed on post-mortem brain tissues, some using quantitative (biochemical) and some qualitative (histochemical) assays to demonstrate lipid peroxidation. In this article I will focus my attention only on the quantitative analyses.

Historically, malonaldehyde (MDA) and the thiobarbituric acid-reactive substances (TBARS) assays have been the first and most employed techniques to biochemicallly quantitate lipid peroxidation in AD. The majority of these investigations have shown higher MDA and/or TBARS levels in AD than in control brains (Subbarao et al. 1990; Balazs and Leon 1994). Several studies have attempted to measure serum and plasma levels of MDA in living patients with AD, but conflicting results have been reported (Jeandel et al. 1989; Bourdel-Marchyasson et al. 2001). Another by-product of the oxidation of PUFA that has been quantified is 4-hydroxy-2-nonenal (4-HNE). Its levels have been reported to be elevated in AD brain tissue as well as in ventricular cerebrospinal fluid (CSF) (Markesbery and Lovell 1998; Lovell et al. 1997). However, its use in biological systems has been relatively limited, because its levels are generally low and large quantities of sample are needed to measure it.

Initially, we investigated levels of two distinct F_2-isoprostanes isomers – iPF2_α-III and iP$F_{2\alpha}$-VI – in post-mortem brain tissues from AD patients. We compared them to levels in tissues of patients with other neurological disorders and to neurologically healthy controls (Praticò et al. 1998). We found that the levels of these two isomers were markedly higher in both frontal and temporal poles, as well as in the ventricular CSF of AD subjects than in the other groups. Remarkably, no such difference was observed in the cerebellum, an area traditionally devoid of the pathological hallmarks of the disease. This study confirmed that oxidative stress is a feature of AD, and that it localizes in areas specifically affected by the disease.

However, the study did not address the question of whether or not oxidative imbalance and subsequent lipid peroxidation are early components or final common steps of the neurodegenerative process in AD. For this reason, we collected urine, plasma and CSF from subjects with a clinical diagnosis of AD and from age-matched controls (Praticò et al. 2000a). We found that, compared with controls, AD patients had increased CSF, plasma and urinary levels of a major F_2-isoprostane, 8,12-iso-iP$F_{2\alpha}$-VI. Urinary and circulating plasma levels of this isoprostane directly correlated with the levels in CSF of AD patients, suggesting a common mechanism of formation: brain oxidative stress. Interestingly, we observed a direct correlation between CSF 8,12-iso-iP$F_{2\alpha}$-VI levels and CSF tau and an inverse correlation with the percentage of CSF Aβ1-42 in AD patients. Furthermore, we found a significant correlation between 8,12-iso-iP$F_{2\alpha}$-VI CSF levels and the severity of the dementia in AD patients, as measured by two of the most common cognitive tests, the mini-mental state examination and the dementia severity rating scale (Praticò et al. 2000a).

Taken together, these findings suggest that elevation of this isoprostane not only reflects brain oxidative stress but also correlates with the progression of the disease. To further confirm this observation, we assayed plasma levels of several endogenous anti-oxidants in these patients. We found that, while there was no difference between AD patients and controls for vitamin A, uric acid and β-carotene levels, a significant difference was observed for vitamin C, E, lycopene and α-carotene levels (Praticò and Sung 2004). Our studies support the hypothesis that oxidative imbalance and subsequent lipid peroxidation occur early in the course of this dementing disorder, thereby implicating them as potential contributors to brain degeneration in AD. Furthermore, the fact that urine levels correlated with CSF levels offers, for the first time, the potential for using a non-invasive tool to investigate brain oxidative damage and to monitor therapeutic responses in AD. These findings were recently confirmed by two different groups (Tuppo et al. 2001; Kim et al. 2004).

To further confirm that brain oxidative imbalance is an early event in AD, we assayed levels of 8,12-iso-iPF$_{2\alpha}$-VI in young patients with Down's syndrome. It is known that these subjects exhibit increased concentration of Aβ in the brain, with a precocious AD-like pathology and dementia later in life (Coyle et al. 1986). We found elevated 8,12-iso-iPF$_{2\alpha}$-VI levels in urine samples of subjects with Down's syndrome compared with those of matched controls, which correlated with the duration of the disease (Praticò et al. 2000b).

AD has a long stage of neuropathological changes and cognitive decline before it is diagnosed. In recent years, it has been shown that the onset of AD is preceded by an interim phase known as mild cognitive impairment (MCI). Despite the potential heterogeneity of an MCI diagnosis, some studies have suggested that this condition is associated with up to a 50% probability of progressing to symptomatic AD within a four-year period (Petersen et al. 2001). Since individuals with MCI are felt to be at high risk to progress to a clinical diagnosis of AD, we have investigated if they, similar to AD patients, have signs of brain lipid peroxidation and oxidative stress. For this purpose we compared the levels of 8,12-iso-iPF$_{2\alpha}$-VI in urine, plasma and CSF in AD and MCI subjects and age-matched controls (Praticò et al. 2002). First, we confirmed that the AD patients had the highest levels of this marker among the three groups. However, we found that the patients who met standardized clinical criteria for MCI also had increased CSF, plasma and urinary levels of 8,12-iso-iPF$_{2\alpha}$-VI compared with cognitively normal elderly controls. In accordance with previous reports, we found that, among the three groups studied, AD patients had the highest values for CSF tau and the lowest percentage of CSF Aβ1-42 (Andreasen et al. 2001). No significant difference was observed between MCI subjects and age-matched controls for either CSF marker. By contrast, MCI subjects had CSF 8,12-iso-iPF$_{2\alpha}$-VI levels that were significantly higher than those in controls. Taking into account that CSF tau and the percentage of CSF Aβ1-42 levels are considered to be good markers of AD neuropathology and disease progression, this observation would further support the hypothesis that brain oxidative stress is present before the detection of the typical neuropathological changes of early AD (Nanomura et al. 2001). Remarkably, in our study we found that MCI subjects are different from elderly control subjects only with respect to this marker of oxidative stress. This finding suggests that measurement of 8,12-iso-iPF$_{2\alpha}$-VI may provide a reliable biomarker that could help in identifying those MCI subjects who have an early brain oxidative stress. This subset of individuals could be at higher risk for progression to AD and might also be the most responsive to antioxidant treatment. However, future longitudinal studies measuring this marker in MCI patients are needed to test these hypotheses. Supporting this concept, a recent study reported that peripheral levels and activities of several anti-oxidants are lower in MCI and AD subjects as compared to healthy controls (Rinaldi et al. 2003).

Oxidative stress and non-AD dementia

As stated previously, AD is the most frequent cause of neurodegeneration with dementia. However, there is a large group of neurodegenerative diseases with dementia that are histopathologically different from AD and can be grouped under the general term of non-AD dementias. Among them, frontotemporal dementia (FTD) is probably the most representative group.

FTD is a heterogeneous group of neurodegenerative conditions that accounts for 3 to 10% of all dementia. Typically, all FTDs are characterized by a striking and profound degeneration of the frontal and temporal lobe, similar to AD (McKhann et al. 2001). However, while these brain regions are affected in most FTDs, several FTD subtypes have been proposed, depending on the prevalent pathological involvement of other cortical, subcortical, white matter and brain stem regions (Mann et al. 1993). One of the most common sporadic FTD subtypes is known as dementia lacking distinctive histopathology (DLDH), but other FTD groups include progressive supranuclear palsy (PSP), FTD with parkinsonism linked to chromosome 17 (FTD-17), and Pick's disease (Arnold 2000). Despite this phenotypic heterogeneity, there are reports showing evidence of oxidative stress in some cases of FTDs (Castellani et al. 1995; Gerst et al. 1999), suggesting that, similar to AD, brain oxidative stress may be a common disease pathway or a final common step of the neurodegenerative process.

Even though AD and FTD are both neurodegenerative dementias, their neuropathologies are significantly different. For instance, abundant neurofibrillary tangles and amyloid plaques are diagnostic for AD, whereas some FTD variants have disease-specific tau lesions (e.g., Pick's bodies) and others are not associated with tau or any other specific lesions (DLDH). All FTDs present several common aspects of all neurodegenerative disorders (e.g., neuronal loss, reactive astrocytosis, and microglia proliferation) that are considered to be common responses of the brain to different insults, such as oxidative stress and inflammation. However, conflicting results have been reported regarding the notion that brain oxidative stress also plays a role in most of the FTDs. A possible explanation for these contrasting findings is that some of these studies used markers of lipid peroxidation that are known to lack specificity and sensitivity, such as MDA and TBARS assays (Albers et al. 2000). For example, a study performed with PSP brain tissues reported immunocytochemical and biochemical evidence for lipid peroxidation, but not protein oxidation or glycoxidation (Odetti et al. 2000). By contrast, another group using an antiserum against 4-HNE pyrrole adducts failed to show any positive immunostaining in PSP, and only a minimal reaction in one case of Pick's disease (Montine et al. 1997).

To investigate whether an increase in brain lipid peroxidation is part of a generic response to neurodegeneration, we recently compared levels of 8,12-iso-$iPF_{2\alpha}$-VI in different brain regions from patients assigned a post-mortem diagnosis of FTDs, or AD or age-matched normal elderly. In this study, we demonstrated that brain levels of 8,12-iso-$iPF_{2\alpha}$-VI were significantly increased in AD but not in any of the FTD subtypes considered, and no difference in the levels of this marker was found between FTD and controls. This finding was true when we analyzed each FTD subtype alone, and also when we divided all of the subgroups into cases with and without tau deposition and abnormalities (Yao et al. 2003). We extended this observation in ventricular CSF of histopathologically confirmed cases of FTD as well as in CSF of living patients with a clinical diagnosis of FTD (Grossman et al., submitted for publication). In this study, we found that, when compared with controls, levels of 8,12-iso-iPF$_{2\alpha}$-VI in AD patients were elevated, but this was not the case for FTD subjects. These findings support the hypothesis that oxidative stress to the brain is highly specific to AD, and it is not uniform across neurodegenerative diseases, even among those associated with similar clinical and pathological features, such as FTDs. This finding would suggest that mechanisms of diseases underlying FTDs are not able to induce an increase in brain oxidative stress.

Conclusions

The fact that age is a risk factor for dementia has provided the initial basis for the involvement of oxidative stress in this disease. While an increase in lipid peroxidation has been widely and extensively demonstrated in AD, conflicting results have been reported for non-AD dementias, because for many years our understanding of the role that oxidative stress plays in neurodegeneration has been hampered by the lack of sensitivity and specificity of many of the assays used. However, while some traditional views maintain that oxidative stress is the simple result of neurodegeneration, the contribution of ROS-mediated reactions to neuronal loss in AD and non-AD dementias is now beginning to be elucidated. F_2-Isoprostanes are a new class of lipids derived from the peroxidation of polyunsaturated fatty acids. Consistent data have been accumulated that clearly pinpoint them as a reliable and non-invasive approach for studying oxidative stress in vivo. Today we have consistent evidence to support the concept that oxidative stress to the brain is not a general and final common pathway of the neurodegenerative process, but it may involve specific mechanism(s) characteristic of the disease process that are present in AD but not in other dementias. Thus, oxidative stress is an early event during the evolution of AD and it might play a more important role in its pathogenesis than in the pathogenesis of non-AD dementias. This new concept provides a rational basis for therapeutic intervention with antioxidants in AD subjects at the earliest possible stage of the disease.

Acknowledgments

Part of the work described in this article was supported by grants form the National Institutes of Health (AG-11542, AG-14382, AG-14447).

References

Albers DS, Augood SJ, Park LCH, Browne SE, Martin DM, Adamson J, Hutton M, Standaert DG, Vonsattel JPG, Gibson GE, Beal FM (2000) Frontal lobe dysfunction in progressive

supranuclear palsy: evidence for oxidative stress and mithocondrial impairment. J Neurochem 72:878–881.
Andreasen N, Minthon L, Davidsson P, Vanmechelen E, Vanderstichele H, Winblad B, Blennow K (2001) Evaluation of CSF-tau and CSF-Aβ42 as diagnostic markers for Alzheimer's disease. Arch Neurol 58:373–379.
Arnold SE (2000) Cellular and molecular pathology of the frontal dementias. In: Clark CVM, Trojanowski JQ (eds.) Neurodegenerative dementias: clinical features and pathological mechanisms. New York, NY; McGraw-Hill, pp..291–301.
Balazs L, Leon M (1994) Evidence of an oxidative challenge in the Alzheimer's brain. Neurochem Res 19: 1131–1137.
Bourdel-Marchyasson I, Delmas-Beauvieux MC, Peuchant E, Richard-Harston S, Decamps A, Reigner B, Emeriau JP, Rainfray M (2001) Antioxidant defenses and oxidative markers in erythrocytes and plasma from normally nourished elderly Alzheimer patients. Age Ageing 30:235–241.
Castellani RJ, Smith MA, Richey PL, Kalaria R, Gambetti P, Perry G (1995) Evidence for oxidative stress in Pick disease and corticobasal degeneration. Brain Res 696:268–271.
Christen Y (2000) Oxidative stress and Alzheimer's disease. Am J Clin Nutr 71:621S–629S.
Clark CM (2000) Clinical manifestations and diagnostic evaluation of patients with Alzheimer's disease. In: Clark CM, Trojanowski JQ (eds) Neurodegenerative dementias: clinical features and pathological mechanisms. New York, McGraw-Hill, pp. 95–114.
Coyle JT, Oster-Granite ML, Gearhart JD (1986) The neurobiologic consequences of Down syndrome. Brain Res Bull 16:773–787.
Floyd RA (1999). Antioxidants, oxidative stress, and degenerative neurological disorders. Proc Soc Exp Biol Med 222:236–245.
Gerst JL, Siedlak SL, Numonura A, Castellani R, Perry G, Smith MA (1999) Role of oxidative stress in frontotemporal dementia. Dement Geriatr Cogn Disord 10(S):85–87.
Grossman M, Farmer J, Leight S, Work M, Moore P, van Deerlin V, Praticò D, Clark CM, Coslett HB, Chatterjee A, Trojanowski JQ, Lee VM-Y (2004) Cerebrospinal fluid tau is reduced in frontotemporal dementia and corticobasal degeneration. Ann Neurol, submitted.
Halliwell B, Gutteridge JMC (1999). Free radicals in biology and medicine. 3rd Ed. Oxford: Oxford University Press.
Harman D (1956) Aging: a theory based on free radicals and radiation chemistry. J Gerontol 11:298–300.
Jeandel C, Nicolas MB, Dubois F, Nabet-Belleville F, Penin F, Cuny G (1989) Lipid peroxidation and free radical scavengers in Alzheimer's disease. Gerontology 35: 275–282.
Kim KM, Jung BH, Paeng K-J, Kim I, Chung BC (2004) Increased urinary F_2-isoprostanes levels in the patients with Alzheimer's disease. Brain Res Bull 64:47–51.
Lovell MA, Ehmann WD, Mattson MP, Markesbery WR (1997) Elevated 4-hydroxynonenal in ventricular fluid in Alzheimer's disease. Neurobiol Aging 18: 457–461.
Mann DA, South PW, Snowden JS, Neary D (1993) Dementia of the frontal lobe type-neuropathology and immunohistochemistry. J Neurol Neurosurg Psychiat 56: 605–614.
Markesbery WR, Lovell MA (1998) Four-hydroxynonenal, a product of lipid peroxidation, is increased in the brain in Alzheimer's disease. Neurobiol Aging 19: 33–36.
McKhann GM, Grossman M, Albert MS, Trojanowski JQ (2001) Clinical and pathological diagnosis of frontotemporal dementia: report of work group on frontotemporal dementia and Pick's disease. Ann Neurol 58:1803–1809.
Montine KS, Kim PJ, Olson SJ, Markesbery WR, Montine TJ (1997) 4-Hydroxy-2-nonenal pyrrole adducts in human neurodegenerative disease. J Neuropathol Exp Neurol 56:866–871.
Nanomura A, Perry G, Aliev G, Hirai K, Takeda A, Balarj EK, Jones PK, Ghanbari H, Wataya T, Shimohama S, Chiba S, Atwood CS, Petersen RB, Smith MA (2001) Oxidative damage is the earliest event in Alzheimer's disease. J Neuropathol Exp Neurol 60:759–767.

Odetti P, Garibaldi S, Norese R, Angelini G, Marinelli L, Valentini S, Menini S, Traverso N, Zaccheo D, Siedlak S, Perry G, Smith MA, Tabaton M (2000) Lipoperoxidation is selectively involved in progressive supranuclear palsy. J Neuropath Exp Neurol 59:393–397.
Petersen RC, Doody R, Kurz A, Mohs RC, Morris JC, Robins PV, Ritchie K, Rossor M, Thal L, Winbland B (2001) Current concepts in mild cognitive impairment. Arch Neurol 58:1985–1992.
Praticò D (2001) Alzheimer's disease and oxygen radicals: new insights. Biochem Pharmacol 63:563–567.
Praticò D (2002) Lipid peroxidation and the aging process.Sci Aging Knowledge Environ. Dec 18, 2002 (50): re5.
Praticò D, Delanty N (2000) Oxidative injury in Diseases of the Central Nervous System: Focus on Alzheimer's disease. Am J Med 109:577–585.
Praticò D, Sung S (2004) Lipid peroxidation and oxidative imbalance: early functional events in Alzheimer's disease. J Alz Dis 6:171–175.
Praticò D, Lee V M-Y, Trojanowski JQ, Rokach J, FitzGerald GA (1998) Increased F_2-isoprostanes in Alzheimer's disease: evidence of enhanced lipid peroxidation in vivo. FASEB J 12:177–178.
Praticò D, Clark CM, Lee V MY, Trojanowski JQ, Rokach J, FitzGerald GA (2000a) Increased 8,12-*iso*-$iPF_{2\alpha}$-VI in Alzheimer's disease: correlation of a noninvasive index of lipid peroxidation with disease severity. Ann Neurol 48:809–812.
Praticò D, Iuliano L, Amerio G, Rokack J, Sabatino F, Violi F (2000b) Down's syndrome is associated with increased 8,12-*iso*-$iPF_{2\alpha}$-VI levels: evidence for enhanced lipid peroxidation in vivo. Ann Neurol 48:795–798.
Praticò D, Lawson JA, Rokach J, FitzGerald GA (2001) The isoprostanes in biology and medicine. Trends Endocrinol. Metab. 12:243–247.
Praticò D, Clark CM, Liun F, Lee V MY, Trojanowski JQ (2002) Increase of oxidative stress in mild cognitive impairment: a possible predictor of Alzheimer's disease. Arch Neurol 59:972–976.
Rinaldi P, Polidori MC, Metastasio A, Mariani E, Mattioli P, Cherubini A, Catani M, Cecchetti R, Senin U, Mecocci P (2003) Plasma antioxidants are similarly depleted in mild cognitive impairment and in Alzheimer's disease. Neurobiol Aging 24:915–919.
Smith MA, Rottkamp CA, Nunomura A, Raina AK, Perry G (2000) Oxidative stress in Alzheimer's disease. Biochim Biophys Acta 1502:139–144.
Subbarao KV, Richardson JS, Ang LC (1990) Autopsy samples of Alzheimer's cortex show increased peroxidation in vitro. J Neurochem 55: 342–345.
Tuppo EE, Forman LJ, Spur BW, Chan-Ting RE, Chopra A, Cavalieri TA (2001) Sign of lipid peroxidation as measured in the urine of patients with probable Alzheimer's disease. Brain Res Bull 54:565–568.
Yao Y, Zhukareva V, Clark CM, Rokach J, Lee VM-Y, Trojanowski JQ, Praticò D (2003) Enhanced brain levels of 8,12-*iso*-$iPF_{2\alpha}$-VI differentiate AD from frontotemporal dementia. Neurology 61:475–478.

Early Onset Familial Alzheimer's Disease: Is a Mutation Predictive of Pathology?

Douglas Galasko[1]

Summary

In patients with familial Alzheimer's disease (FAD) with early onset, mutations have been identified in three genes: Presenilin 1 (PS1), Presenilin 2 (PS2) and amyloid protein precursor (APP). Although many different mutations have been recorded for each gene, in most cases the clinical picture is typical of AD, with earlier onset than in sporadic AD. The question of whether mutations in these genes invariably predict AD pathology is the subject of this review. Almost all mutations in APP, PS1 and PS2 promote the development of neuropathological findings of AD, with deposition of amyloid beta protein (Aβ) to form prominent amyloid pathology. Progress has been made in correlating the effects of mutations on species of Aβ with vascular versus parenchymal deposition of Aβ. There are some exceptions to what appears to be an otherwise inevitable relationship. First, a few innocent mutations in APP and PS1 have been described. Second, in rare families with PS1 mutations, there are mutation-bearing individuals who may have escaped clinical disease. Third, some APP mutations alter the sequence of Aβ and lead to severe angiopathy with hemorrhage, rather than the plaques and tangles of AD. Finally, some families with fronto-temporal dementia (FTD) have shown PS1 mutations. Although neuropathologic studies are limited at present, the FTD association raises questions about whether amyloidogenic disease pathways are the only mechanisms that lead to neuropathological changes in early onset AD with presinilin (PS) mutations.

Introduction

In over 95% of patients with Alzheimer's disease (AD), the onset of dementia occurs in late life, after age 60, and the prevalence rises exponentially at higher ages. The brain pathology that defines AD includes deposition of amyloid to form senile plaques (SP), loss of synapses and neurons, and degenerative changes within neuronal processes (neurites) and within neuronal cell bodies [neurofibrillary tangles (NFT)]. There are also variable degrees of vascular amyloid deposits [amyloid angiopathy (AA); reviewed in Dickson 2001]. A common structural theme in AD

[1] Professor, Department of Neurosciences, UCSD, and Staff Physician, Veterans Affairs Medical Center, 3350 La Jolla Village Drive, San Diego, CA 92161, USA.

Cummings et al.
Genotype – Protéotype – Phenotype
Relationships in Neurodegenerative Diseases

lesions is aggregation of molecules: beta-amyloid protein (Aβ) is deposited in SP, and the microtubule-associated protein tau forms aggregates in neurites and in NFT. Some patients also develop aggregates of α-synuclein, which form neurites and Lewy bodies. These are found in the amygdala in about 40% of patients with AD and in a wider distribution with involvement of the neocortex in about 20% of patients. Early onset AD refers to an important minority of patients with onset arbitrarily defined as before age 60. Familial cases of early onset AD are typically inherited in an autosomal dominant pattern and are found to be associated with mutations in three genes: amyloid precursor protein (APP; Goate et al. 1991), Presenilin 1 (PS1; Sherrington et al. 1995), and Presenilin 2 (PS2; Levy-Lehad et al. 1995a). Many new mutations in these genes have been subsequently identified. This review will examine the general question of whether pathology consistent with AD is inevitable in people who bear mutations in APP, PS1 or PS2. A more nuanced question is whether knowledge of the specific mutations involved can help predict age at onset of dementia, variations in the clinical syndrome, and the neuropathological picture.

APP Processing, the Presenilins, and Aβ Protein

The proteins encoded by the genes for APP, PS1 and PS2 interact along pathways of the cellular processing of APP (reviewed in Annaert and De Strooper 2003, and see Fig. 1). APP is a Type I transmembrane protein, with three main isoforms consisting of 695, 751 and 770 amino acids. APP is enriched in the brain and undergoes sequential proteolytic cleavage that produces secreted forms of APP, C-terminal stubs, and Aβ, a 37–43 amino acid protein. Aβ is contained within the sequence of APP and is released after APP is cleaved by the enzyme β-secretase followed by γ-secretase (Vassar and Citron 2000). Aβ is normally produced in the

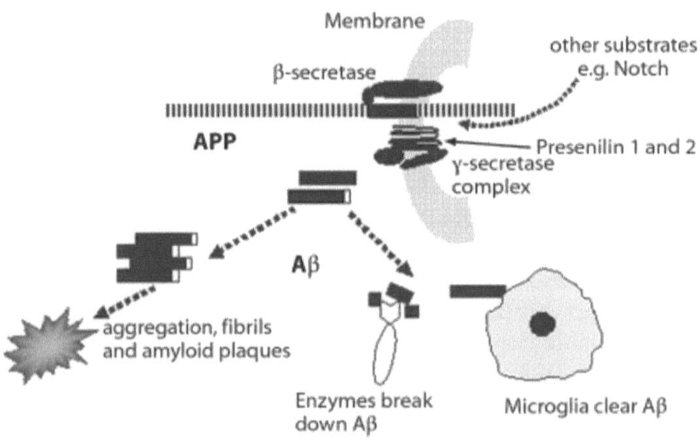

Fig. 1. Schema of APP processing by β-amyloid secretase and γ-secretase (containing the Presenilins), giving rise to Aβ. The figure illustrates the production of different lengths of Aβ from APP. Longer forms of Aβ, particularly Aβ42, are more prone to aggregate.

brain and is rapidly cleared, with a relatively short half-life. In AD, the balance between production and clearance is altered, so that Aβ aggregates and becomes deposited in the brain. This process results in the formation of SP, which are thought to precede NFT and other changes. It is not clear which forms or species of Aβ are the most important in AD pathogenesis. There is evidence to support pathogenic roles for soluble Aβ, oligomers ranging from 2–12 molecules of Aβ, fibrils and plaques, and a role for intracellular Aβ is difficult to exclude. Recent experimental data favor oligomers as being potentially toxic to synapses and neurons (Klein et al. 2001; Walsh et al. 2002).

PS1 and PS2 are integral membrane proteins that are predicted to span the membrane 6–8 times. The PS form part of the multiprotein γ-secretase complex, and appear to contain the active site of the enzyme, characterized by two aspartyl residues in PS. γ-Secretase cleaves a number of proteins other than APP. The cleavage of APP results in a series of species of Aβ of varying lengths at the C-terminus and includes Aβ peptides ending at amino acids 37, 38, 40, 42 or 43. Most of Aβ ends at amino acid 40, whereas about 10–15% ends at amino acid 42. Aβ42 aggregates more readily than Aβ40, shows relatively greater in vitro toxicity, and is the predominant form of Aβ deposited in plaques in AD (Hardy and Selkoe 2002).

Reported Mutations in APP and the Presenilins

To date, over 10 mutations have been reported in APP (Table 1). Many of these flank the Aβsequence, which alters APP processing and Aβ production. Other mutations are within the Aβ sequence and lead to the production of forms of Aβ with altered amino acid composition, which alters the solubility Aβ or its potential to aggregate. Over 100 mutations have been reported in PS1, located throughout the length of this 467 amino acid protein. Some examples are listed in Table 2 and Table 3. At least nine pathogenic PS2 mutations have been reported (well-characterized PS2 mutations are shown in Table 3). FAD associated with these genes has an age at onset of generally before 60, and often markedly earlier. PS1 mutations account for the majority of cases of FAD and have a mean age of onset in the early 40s. The youngest documented patients have onset recorded in their early 20s. In patients with APP mutations, the mean age at onset is about 50. PS2 mutations show the greatest heterogeneity, with mean onset in the late 50s, but a wide range exists within and between families. For an extensive list of specific mutations, see www.alzforum.org or www.uia.ac.be/ADMutations.

The penetrance of mutations in these three genes is extremely high. However, there are rare PS1 pedigrees that include mutation-bearers who have lived beyond age 60 without developing AD (Rossor et al. 1996, Poorkaj et al. 1998). APP mutations are similarly almost always associated with early onset AD. In PS2 families, onset after age 65 sometimes occurs, which raises the possibility of coincident sporadic AD in older family members. Age at onset is an imprecise measure in AD. Within families, the onset of AD can vary, at times dramatically, even though people harbor the same mutations; differences in age at onset of as much as 20 years have been noted (Lopera et al. 1997). The Apolipoprotein E gene (APOE)

Table 1. APP mutations associated with early onset FAD, with reported neuropathology and/or mechanism involving Aβ.

Mutation	Clinico-pathologic phenotype	Age at onset (Mean or range)	Effect on Aβ	Reference
LysMet670/ AsnLeu "Swedish"	AD ->→ SP, NFT, mild AA	44–59	↑ Aβ40 and 42 Better substrate for BACE	Mullan et al. 1992, Bogdanovic, 2002
His677Arg	AD ->→ SP, NFT	55		Janssen et al. 2003
D678N	AD. ? pathology	60		Wakutani et al. 2004
Ala692Gly "Flemish"	Dementia, hemorrhagic strokes. CAA + large plaques	40–60	↑ Aβ40 and 42	Cras et al. 1998
Glu693Gly "Arctic"	Dementia, hemorrhagic strokes. CAA, large plaques	57	slight ↓ Aβ The mutant Aβ forms ↑ protofibrils	Nilsberth et al. 2001
Glu693Gln "Dutch"	Dementia, hemorrhagic strokes. CAA, hemorrhage	± 50	↓ plasma Aβ42 ↑ oligomers and aggregation of Aβ	Bornebroek et al. 2003
Glu693Lys "Italian"	Brain hemorrhage, ? pathology	?	?	
Asp694Asn	Brain hemorrhage "Iowa" CAA, SP with mainly Aβ40, NFT	56–70 or AD.	↑ aggregation	Grabowski et al. of Aβ 2001
Thr714Ile	AD. Diffuse plaques, almost all Aβ42, N-terminal truncated, non-fibrillar	34	↑↑ Aβ42 production	Kumar-Singhw et al. 2000
Val715Met (French)	AD. ? pathology	40–60	↓Aβ40, ↑ N-terminally truncated Aβ42	Ancolio et al. 1999
Ile716Val (Florida)	AD. ? pathology	55	↑ Aβ42/Aβ40	Eckman et al. 1997
Val717Leu (Indiana)	AD. ? pathology	38	↑ Aβ42/Aβ40	Murell et al. 1991
Val717Ile (London)	AD. ? pathology	50–60	↑ Aβ42/Aβ40	Goate et al. 1991
Leu723Pro	AD. ? pathology	45–60	slight ↑ Aβ42/Aβ40	Kwok et al. 2000

Not shown: mutations that are non-pathogenic or of dubious pathogenic status: Glu665Asp, Ala673Thr, Ala713Thr, Ala713Val. Abbreviations: AD = Alzheimer's Disease, CAA = Cerebral Amyloid Angiopathy, SP = senile plaques, NFT = neurofibrillary tangles, AA = amyloid angiopathy, BACE = beta amyloid cleavage enzyme, Aβ= amyloidβ -protein, ? pathology = unknown or unreported neuropathology, ↑ = increase, ↓ = decrease

Table 2. Selected PS1 mutations associated with early onset FAD, with unusual clinical phenotypes

Mutation	Clinico-pathologic phenotype	Age at onset (Mean or range)	Effect on Aβ	Reference
AD + Parkinsonism/Lewy bodies				
M233V	AD + Parkinsonism, seizures, rapid course → AD + DLBD	32	↑ Aβ42	Houlden et al. 2001
Glu184Asp	AD + Parkinsonism, hallucinations. AD + DLBD	40		Yokota et al. 2002
AD + Spastic paraparesis Δ exon 9 Finnish; Australian	AD, spasticity, increased reflexes, poor gait Cotton wool plaques	20–61	↑ Aβ42	Crook et al.1998, Kwok et al. 1997
ΔIle83/ ΔMet84 Scottish	AD, leg spasticity Cotton wool plaques	42–28	↑ Aβ42	Steiner et al. 2001
Arg278Thr	AD, spastic paraparesis.		↑ Aβ42	Kwok et al. 1997
Leu271Val (Δ exon 8)	AD, spasticity Cotton wool plaques	22–35	↑ Aβ42	Kwok et al. 2003
insIle157	AD plus spastic paraparesis, dystonia and dysarthria. Plaques and tangles, widespread	28–33		Moretti et al. 2004
Fronto-temporal dementia				
Leu113Pro	Personality and behavior changes. Preserved memory. ? pathology	38–50		Raux et al. 2000
insArg352	Personality and behavior changes. ? pathology	56	Plasma Aβ unchanged ↓ processing of notch	Amtul et al. 2002
Gly183Val	Antisocial behavior, apathy, withdrawal, disinhibition. Pick bodies.	52	↓ Aβ Altered splicing of PS1 with ? ? (is this clear?)loss of function	Dermaut et al 2004

DLBD = diffuse Lewy body disease, ? pathology = unknown or unreported neuropathology.

Table 3. Selected PS1 and PS2 mutations associated with early onset FAD, with typical clinical features and reported neuropathology and/or mechanism involving Aβ.

Mutation	Clinico-pathologic phenotype	Age at onset (Mean or range)	Effect on Aβ	Reference
Presenilin 1 – typical AD phenotype (selected from over 100 reported families with mutations for cases with well-studied pathology)				
Δ4 (deletion at an intron 4 splice-donor site)	AD. Plaques and NFT	40	↑ A β42, most likely due to a form of PS1 with a Thr113-114 insert	Tysoe et al. 1998
Pro117Ser	AD. Plaques, many tangles and severe neuritic change	24–28 4	↑ Aβ42	Dowjat et al. 200
Glu120Asp	AD. Plaques and NFT	34–53	↑ Aβ42/40 ratio	Bird et al. 1989
Asn135Asp	AD. Plaques and very many NFT		30–35	Crook et al. 1997
Ala260Val	AD. Plaques and NFT	36–54	↑ Aβ42	Bird et al. 1989
Glu280Ala Colombian	AD. Plaques and NFT	34–62	↑ Aβ42/40 ratio	Lopera et al. 1997
Leu282Val	AD. CAA with surrounding NFT.	41–53	↑ Aβ42/40 ratio	Cruts et al. 2001 Dermaut 2001
Gly384Ala Mann et al. 2001	AD. Plaques and NFT, marked CAA		36	↑ Aβ42/40 ratio
Presenilin 2				
Arg62His	AD			Cruts et al. 1998
Asn141Ile "Volga German"	AD. Plaques, tangles, variable AA	50–70	↑ Aβ42/40 ratio	Levy-Lehad et al. 1995
Met239Val	AD. Plaques, NFT, ectopic neurons in subcortical white matter	50–70	↑ Aβ42/40 ratio	Marcon et al. 2004

There are over 100 reported mutations in PS1 and 11 reported mutations in PS-2. Websites that catalog these are mentioned in the text.

modifies the age at onset of sporadic AD. APOE has been difficult to study in patients with FAD, because of the relatively few affected individuals from any particular family (Levy-Lahad et al. 1995b). In a large extended kindred with AD associated with the PS1 E280A mutation, patients with AD who also had an APOE ε4 allele had younger age at onset than patients without an ε4 allele (Pastor et al. 2003).

Finding a mutation in APP, PS1 or PS2 in an individual patient with early onset AD is strongly suggestive that the mutation is pathogenic. However, there are benign amino acid substitutions within these proteins that do not appear to alter their function and are not associated with an increased risk of AD (Lleo et al. 2002). Stronger evidence that a novel mutation in any of these genes is causal can be provided by neuropathologic examination of brain tissue from an affected person to verify AD pathology, and by further DNA studies to demonstrate that the mutation segregates with a clinical picture of dementia in affected members of a family and is not found in healthy older individuals in the population at large. This evidence is not always available, e.g., in small families or when parents died at an early age. Thus, in a single affected patient, the predictive value of a novel mutation, although high, will be influenced by factors such as the age at onset. A positive family history increases the yield of finding mutations in patients with early onset AD (Janssen et al. 2003).

Clinicopathological Phenotypes in FAD Mutations

Typical clinical and pathological findings

As in sporadic AD, FAD is associated with the gradual onset and progression of memory impairment, followed by impairment of abilities such as language, executive function, calculation, and other abilities, leading to impaired daily function. Myoclonus and seizures occur to a varying extent in late-onset AD but are slightly more common in early onset FAD (Bird et al. 1989; Mann et al. 2001). The duration of dementia from onset to death varies widely in FAD, analogous to late life AD, where a range of three to 20 years has been reported. On average, the duration from onset to death might be expected to be longer in dementia with early onset, because younger patients have better general medical health. The fact that FAD has a similar or shorter duration than late onset AD suggests that the build-up and progression of pathology is accelerated in FAD. Comparisons between families bearing mutations in PS1, PS2 and APP suggest that PS1 mutations are usually the most malignant. Patients with PS2 mutations tend to have onset later in life, with longer disease duration and increased variability of disease expression. APP mutations lead to intermediate ages of onset.

The pathology of PS1 mutations has been extensively studied and compared with that of late onset AD (Gomez-Isla et al. 1999; Mann et al. 2001). There appears to be an overall increase in Aβ deposition in PS mutation-bearers relative to late onset AD, but the same regions of the brain are affected. Immunostaining with antibodies selective for different species of Aβ shows that the increase is primarily due to Aβ42, not Aβ40. This finding is consistent with studies of cells transfected with mutant forms of PS1, which typically show relative overproduction of Aβ42 relative to cells transfected with wild type PS1. There is a striking correlation between the extent of Aβ deposition in the brain and the levels of secretion of Aβ42 in cell lines transfected with a range of PS1 mutations. Two broad patterns of Aβ pathology have been described in PS1 mutations: a type 1 pattern is more likely for mutations in the first 200 codons of PS1, and a type 2 pattern for mutations after codon 200 (Mann et al. 2001). Type 1 shows many diffuse plaques, few cored plaques and mild or moderate amyloid angiopathy, usually limited to leptomeningeal vessels. This pattern corresponds to a marked predominance of deposition of Aβ42 versus Aβ40 and resembles the pattern found in sporadic AD. Type 2 shows relatively fewer diffuse plaques and relatively more abundant and larger cored plaques that often surround blood vessels with severe amyloid angiopathy. This pattern corresponds to deposition of Aβ40 as well as Aβ42. The reasons for greater amyloid angiopathy in the type 2 pattern are not clear. Type 1 is associated with a slightly earlier age at onset and shorter disease duration than type 2.

Pathological changes in PS2 mutations generally include plaques and tangles, with similar immunostaining patterns using antibodies selective for Aβ40 and Aβ42 as described for PS1 (Gomez-Isla et al. 1999). However, in the large Volga German family with a PS2 Asn141Ile mutation, several affected members had severe amyloid angiopathy and cerebral hemorrhage (Nochlin et al. 1998). Factors such as age and modifying genes such as APOE may influence the pathologic phenotype of patients with mutations in PS2.

Neurofibrillary tangles are prominent in patients with PS mutations, and in some patients they exceed the numbers of tangles found in sporadic AD. Additional neuronal changes can be found in early onset FAD. Lewy bodies and neurites containing aggregates of insoluble α-synuclein are found in the amygdala in about 20% of patients with PS or APP mutations (Lippa et al. 1998). It is not clear whether they contribute to the clinical picture, because they are usually restricted to the amygdala. Patients with PS or APP mutations sometimes show clinical features of prominent Parkinsonism or visual hallucinations, which can be associated with more widespread Lewy body formation (e.g., Houlden et al. 2001; Yokota et al. 2002). Other less common findings include vacuolar change (Met146Val and Glu280Ala), and ubiquitin-positive inclusions in the dentate gyrus (Ala260Val).

Atypical clinical pictures and underlying neuropathology

As more well-characterized families are screened for mutations in APP and PS, a more complex clinical spectrum, with phenotypic heterogeneity, is emerging. In several families with PS1 alterations, including splice acceptor site mutations, deletions involving exon 9 or exon 4, and a few missense mutations, the clinical picture is AD plus spastic paraparesis (Kwok et al. 1997; Crook et al. 1998; Steiner et al. 2001; Brooks et al. 2003). The paraparesis often precedes early onset dementia in these families by as much as 5 years. Clinical features of the dementia are typical, with an initial decline of memory followed by impairment in other cognitive abilities. In a recently reported family associated with a novel insert in PS1 (insIle157), affected individuals showed early onset of dementia, with rapid clinical progression, and additional clinical features of spastic paraparesis, dysarthria, limb apraxia and dystonia (Moretti et al. 2004). The combination of dementia and spastic paraparesis may occur in several genetic disorders, including familial British Dementia, some forms of Hereditary Spastic Paraparesis, and Gerstmann-Strausser-Schenkler syndrome (a prion disorder). In families with AD plus spastic paraparesis, "cotton wool" plaques are prominent in the brain (Crook et al. 1998; Brooks et al. 2003). These are large uncored plaques, almost exclusively containing Aβ42 with very little Aβ40, and neuritic responses around the plaque are minimal or lacking. Additional pathology includes degeneration of the lateral corticospinal tracts. In the family with AD plus spastic paraparesis, dystonia and dysarthria (PS1 mutation with insIle157), neuropathologic findings were amyloid plaques and tangles with a more widespread distribution than that typical of AD, including involvement of the motor cortex and basal ganglia.

In patients with APP mutations within the Aβ sequence, the clinical picture is often that of hemorrhagic stroke with dementia and death, as described in Dutch, Flemish, Italian, Swedish and Iowa families (Cras et al. 1998; Grabowski et al. 2001; Nilsberth et al. 2001). The age at onset is in the fifth or sometimes sixth decade. Vascular amyloid deposition predominates, leading to progressive dementia, often with stroke due to cerebral hemorrhage. The hemorrhages vary from microscopic to large lobar hemorrhages and can lead to stepwise progression of dementia, as well as to death from a large hemorrhage. Cerebral amyloid angiopathy (CAA) is the predominant pathological finding. Widespread deposition of amyloid is seen in the walls of leptomeningeal and cortical arterioles, with evidence of hemorrhage, often in a lobar distribution, and white matter pallor is frequently found. Some patients have varying degrees of parenchymal plaques and tangles, but the burden of these typical AD lesions is generally quite low (Cras et al. 1998; Grabowski et al. 2001; Nilsberth et al. 2001). Inherited CAA can also occur in mutations of other genes such as cystatin and transthyretin.

Finally, a few families have recently been reported with a clinical syndrome of fronto-temporal dementia, with prominent changes in personality and behavior such as anti-social acts and disinhibition, in whom novel PS1 mutations were identified. There is only a single brain autopsy from an individual in one family to date. The brain showed frontal (lobar) atrophy, with neuronal loss, Pick bodies, and the absence of plaques (Dermaut et al. 2004).

Mechanisms of disease

The large number of mutations in PS1, PS2 and APP do not include any that are out of frame or that result in a markedly dysfunctional protein. The most parsimonious explanation of how the mutations lead to the early onset of FAD is gain of function, in a way that promotes the aggregation or deposition of Aβ. Studies of plasma from patients with APP, PS1 and PS2 mutations have almost all shown an absolute increase in the concentration of Aβ42 (often accompanied by an increase in Aβ40) or a relative increase in the percentage of Aβ42 to Aβ40 (Scheuner et al. 1996). These findings have been recapitulated by studies of conditioned medium of cultured cell lines transfected with mutant forms of PS or APP (Murayama et al. 1999; reviewed in Lleo et al. 2004). The increased Aβ42 levels promote aggregation, deposition and toxicity at an earlier age than in sporadic AD. Deposition of Aβ40 is likely to occur only once there has been considerable seeding of plaques by Aβ42, but Aβ40 predominates in amyloid angiopathy. Patients with PS1 mutations associated with cotton wool plaques show evidence of increased Aβ42 production (Houlden et al. 2000). Interestingly, in the exon 9 deletion associated with spastic paraparesis and dementia, the deletion creates a new point mutation (Ser290Cys) at the new exon8/10 splice junction, which is responsible for altering PS function in a way that markedly increases production of Aβ42 (Steiner et al. 1999).

In APP mutations that alter the sequence of Aβ and are associated with CAA, only the Flemish (Ala692Gly) mutation leads to an increased concentration of A40 and Aβ42. In the other mutations, levels of Aβ40 and 42 are normal or decreased, but the abnormal forms of Aβ have accelerated rates of fibril formation (Nilsberth et al. 2001). Why this favors vascular deposition of amyloid is not clear.

In a few PS mutations, there appears to be loss of PS function as measured by the processing of other substrates of PS, or by assays that examine whether transfection with human PS can rescue mutant Drosophila deficient in the activity of a PS homolog. One of the key substrates of γ-secretase is Notch, a protein with important roles in determining cell fate during development but which is also expressed in adult life in the brain. It is not known whether changes in Notch processing contribute to any of the neuropathology found in patients with PS mutations. At least one of the PS1 mutations associated with a FTD-like clinical picture does not increase plasma Aβ42 (Dermaut et al. 2004). It is possible that PS1 mutations associated with FTD could operate through mechanisms related to loss of γ–secretase function (Amtul et al. 2002).

Finally, many other deficits have been reported in animal and cellular studies using mutant PS. For example, transgenic mouse models that express mutant PS1 show abnormal calcium signaling (Leissring et al. 2000) and sensitivity to excitotoxicity and other stresses (Guo et al. 1999), which may increase the vulnerability of neurons to damage or degeneration. In the hippocampus of adult mice that overexpressed a PS1 mutant, the survival of newly generated neural progenitor cells was impaired (Wen et al. 2004). There is evidence for neuronal lysosomal pathology in brains of humans with PS mutations and in transgenic mice that overexpress mutant PS (Cataldo et al. 2004).

Implications and Conclusions

The large number of well-described families with early onset AD and pathogenic mutations, and the pathological characterization of many affected individuals, supports the general conclusion that there is an extremely strong relationship between the presence of these mutations and the early development of AD pathology. A number of clinical scenarios arise:

1) In a patient with a typical clinical picture of AD, with age of onset below 60, finding a known pathogenic mutation in PS1, PS2 or APP carries the almost certain prediction of AD pathology. This is likely to be true even if the family does not have multiple affected individuals available for clinical assessment. Documenting a positive family history, and finding that the mutation segre-

gates with disease in the family of interest and does not occur in asymptomatic individuals in the wider population, helps to strengthen the case.

2) In an individual with typical early onset AD, without affected first-degree relatives, finding a novel mutation may require more careful scrutiny. For example, typical AD with onset in the sixth decade could be due to the APOE ε4 allele, especially if the individual is homozygous for ε4. While novel mutations in PS1, PS2 or APP are likely to be pathogenic, there are a few non-pathogenic mutations in PS1 and APP. The topography of the site of the mutation in PS1 (pathogenic mutations appear to involve amino acid substitutions that are located along predicted faces of transmembrane alpha helices; Crook et al. 1997), and the nature of the change in the altered amino acid provide useful information. Measuring plasma Aβ40 and Aβ42 may be helpful, particularly if plasma samples can be obtained from unaffected first-degree relatives. A higher level of certainty can be assigned if autopsy brain tissue becomes available.
3) Unusual syndromes make clinical-pathological predictions less certain. Familial spastic paraparesis with dementia could be due to mutations in genes other than PS1. Because of the rarity of this syndrome as part of FAD, putative novel mutations in PS1 should be supported by additional evidence such as the mutation tracking with affected individuals in a multiplex family or increased plasma Aβ42, to increase the predictive value of finding AD pathology. The recently reported familial FTD kindreds with PS mutations could lead to a radical revision of how these mutations lead to neurodegeneration and disease, but will require more extensive neuropathological correlation.
4) Progress has been made in predicting some of the subtleties of pathological variation in early onset AD, including differences in plaque morphology and varying degrees of associated amyloid angiopathy. The mechanisms that lead to pathological changes in other molecules, leading to NFT and associated neuritic pathology, neuronal death, and the variable presence of synuclein pathology are part of the wider puzzle of AD in general.
5) Within families with the same mutations, there can be heterogeneity regarding age at onset and clinical and pathological features. This finding is consistent with modulation of disease pathways by modifying factors such as other genes and environmental factors.

References

Amtul Z, Lewis PA, Piper S, Crook R, Baker M, Findlay K, Singleton A, Hogg M, Younkin L, Younkin SG, Hardy J, Hutton M, Boeve BF, Tang-Wai D, Golde TE (2002) A presenilin 1 mutation associated with familial frontotemporal dementia inhibits gamma-secretase cleavage of APP and notch. Neurobiol Dis 9:269–273

Ancolio K, Dumanchin C, Barelli H, Warter JM, Brice A, Campion D, Frébourg T, Checler F (1999) Unusual phenotypic alteration of ß amyloid precursor protein (ß-APP) maturation by a new Val-715 → Met βAPP-770 mutation responsible for probable early-onset Alzheimer's disease. Proc Natl Acad Sci USA 96:4119–4124

Annaert W, De Strooper B (2003) A cell biological perspective on Alzheimer's disease. Annu Rev Cell Dev Biol 18:25–51

Bird TD, Sumi SM, Nemens EJ, Nochlin D, Schellenberg G, Lampe TH, Sadovnick A, Chui H, Miner GW, Tinklenberg J (1989) Phenotypic heterogeneity in familial Alzheimer's Disease: a study of 24 knidreds. Ann Neurol 25:12–25

Bogdanovic N, Corder E, Lannfelt L, Winblad B (2002)APOE polymorphism and clinical duration determine regional neuropathology in Swedish APP (670,671) mutation carriers: implications for late-onset Alzheimer's disease. J Cell Mol Med, 6:199–214

Bornebroek M, De Jonghe C, Haan J, Kumar-Singh S, Younkin S, Roos R, Van Broeckhoven C (2003) Hereditary cerebral hemorrhage with amyloidosis Dutch type (AbetaPP 693): decreased plasma amyloid-beta 42 concentration. Neurobiol Dis 14:619–623

Brooks WS, Kwok JB, Kril JJ, Broe GA, Blumbergs PC, Tannenberg AE, Lamont PJ, Hedges P, Schofield PR (2003) Alzheimer's disease with spastic paraparesis and 'cotton wool' plaques: two pedigrees with PS-1 exon 9 deletions. Brain 126:783–791

Cataldo AM, Peterhoff CM, Schmidt SD, Terio NB, Duff K, Beard M, Mathews PM, Nixon RA (2004) Presenilin mutations in familial Alzheimer disease and transgenic mouse models accelerate neuronal lysosomal pathology. J Neuropathol Exp Neurol 63:821–830

Cras P, van Harskamp F, Hendriks L, Ceuterick C, van Duijn CM, Stefanko SZ, Hofman A, Kros JM, Van Broeckhoven C, Martin JJ (1998) Presenile Alzheimer dementia characterized by amyloid angiopathy and large amyloid core type senile plaques in the APP 692Ala->Gly mutation. Acta Neuropathol (Berl) 96: 253–260

Crook R, Ellis R, Shanks M, Thal LJ, Perez-Tur J, Baker M, Hutton M, Haltia T, Hardy J, Galasko D (1997). Early-onset Alzheimer's disease with a presenilin-1 mutation at the site corresponding to the Volga German presenilin-2 mutation. Ann Neurol 42:124–128

Crook R, Verkkoniemi A, Perez-Tur J, Mehta N, Baker M, Houlden H, Farrer M, Hutton M, Lincoln S, Hardy J, Gwinn K, Somer M, Paetau A, Kalimo H, Ylikoski R, Poyhonen M, Kucera S, Haltia M (1998) A variant of Alzheimer's disease with spastic paraparesis and unusual plaques due to deletion of exon 9 of presenilin 1. Nature Med 4:452–455

Cruts M, van Duijn CM, Backhovens H, Van den Broek M, Wehnert A, Serneels S, Sherrington R, Hutton M, Hardy J, St George-Hyslop PH, Hofman A, Van Broekhoven C(1998) Estimation of the genetic contribution of presenilin-1 and -2 mutations in a population-based study of presenile Alzheimer's disease. Human Mol Genet 7:43–51

Dermaut B, Kumar-Singh S, De Jonghe C, Cruts M, Lofgren A, Lubke U, Cras P, Dom R, De Deyn PP, Martin JJ, Van Broeckhoven C (2001) Cerebral amyloid angiopathy is a pathogenic lesion in Alzheimer's disease due to a novel presenilin 1 mutation. Brain 124:2383–2392.

Dermaut B, Kumar-Singh S, Engelborghs S, Theuns J, Rademakers R, Saerens J, Pickut BA, Peeters K, van den Broeck M, Vennekens K, Claes S, Cruts M, Cras P, Martin JJ, Van Broeckhoven C, De Deyn PP (2004) A novel presenilin 1 mutation associated with Pick's disease but not beta-amyloid plaques. Ann Neurol 55: 617–626

Dickson DW (2001) Neuropathology of Alzheimer's Disease and other dementias. Clinics Geriat Medi 17:209–228

Dowjat WK, Kuchna I, Wisniewski T, Wegiel J (2004) A novel highly pathogenic Alzheimer presenilin-1 mutation in codon 117 (Pro117Ser): Comparison of clinical, neuropathological and cell culture phenotypes of Pro117Leu and Pro117Ser mutations. J Alzheimers Dis 6:31–43

Eckman CB, Mehta ND, Crook R, Perez-tur J, Prihar G, Pfeiffer E, Graff-Radford N, Hinder P, Yager D, Zenk B, Refolo LM, Prada CM, Younkin SG, Hutton M, Hardy J (1997) A new pathogenic mutation in the APP gene (I716V) increases the relative proportion of A beta 42(43). Human Mol Genet 6:2087–2089

Goate A, Chartier-Harlin MC, Mullan M, Brown J, Crawford F, Fidani L, Giuffra L, Haynes A, Irving N, James L (1991) Segregation of a missense mutation in the amyloid precursor protein gene with familial Alzheimer's disease. Nature 349:704–706

Gomez-Isla T, Growdon WB, McNamara MJ, Nochlin D, Bird TD, Arango JC, Lopera F, Kosik KS, Lantos PL, Cairns NJ, Hyman BT (1999) The impact of different presenilin 1 and presenilin 2 mutations on amyloid deposition, neurofibrillary changes and neuronal loss in the

familial Alzheimer's disease brain: evidence for other phenotype-modifying factors. Brain 122:1709–1719
Grabowski TJ, Cho HS, Vonsattel JP, Rebeck GW, Greenberg SM (2001) Novel amyloid precursor protein mutation in an Iowa family with dementia and severe cerebral amyloid angiopathy. Ann Neurol 49:697–705
Guo Q, Fu W, Sopher BL, Miller MW, Ware CB, Martin GM, Mattson MP (1999) Increased vulnerability of hippocampal neurons to excitotoxic necrosis in presenilin-1 mutant knock-in mice. Nature Med 5:101–106
Hardy J, Selkoe DJ (2002) The amyloid hypothesis of Alzheimer's disease: progress and problems on the road to therapeutics. 297:353–356
Houlden H, Baker M, McGowan E, Lewis P, Hutton M, Crook R, Wood NW, Kumar-Singh S, Geddes J, Swash M, Scaravilli F, Holton JL, Lashley T, Tomita T, Hashimoto T, Verkkoniemi A, Kalimo H, Somer M, Paetau A, Martin JJ, Van Broeckhoven C, Golde T, Hardy J, Haltia M, Revesz T (2000) Variant Alzheimer's disease with spastic paraparesis and cotton wool plaques is caused by PS-1 mutations that lead to exceptionally high amyloid-beta concentrations. Ann Neurol 48:806–808
Houlden H, Crook R, Dolan RJ, McLaughlin J, Revesz T, Hardy J (2001) A novel presenilin mutation (M233V) causing very early onset Alzheimer's disease with Lewy bodies. Neurosci Lett 313:93–95
Janssen JC, Beck JA, Campbell TA, Dickinson A, Fox NC, Harvey RJ, Houlden H, Rossor MN, Collinge J (2003) Early onset familial Alzheimer's disease: Mutation frequency in 31 families. Neurology 60:235–239
Klein WL, Krafft GA, Finch CE (2001) Targeting small Abeta oligomers: the solution to an Alzheimer's disease conundrum? Trends Neurosci 24:219–224
Kumar-Singh S, De Jonghe C, Cruts M, Kleinert R, Wang R, Mercken M, De Strooper B, Vanderstichele H, Lofgren A, Vanderhoeven I, Backhovens H, Vanmechelen E, Kroisel PM, Van Broeckhoven C (2000) Nonfibrillar diffuse amyloid deposition due to a gamma(42)-secretase site mutation points to an essential role for N-truncated A beta(42) in Alzheimer's disease. Human Mol Genet 9: 2589–2598
Kwok JB, Taddei K, Hallupp M, Fisher C, Brooks WS, Broe GA, Hardy J, Fulham MJ, Nicholson GA, Stell R, St George Hyslop PH, Fraser PE, Kakulas B, Clarnette R, Relkin N, Gandy SE, Schofield PR, Martins RN (1997) Two novel (M233T and R278T) presenilin-1 mutations in early-onset Alzheimer's disease pedigrees and preliminary evidence for association of presenilin-1 mutations with a novel phenotype. Neuroreport 8:1537–1542
Kwok JB, Li QX, Hallupp M, Whyte S, Ames D, Beyreuther K, Masters CL, Schofield PR (2000) Novel Leu723Pro amyloid precursor protein mutation increases amyloid beta42(43) peptide levels and induces apoptosis. Ann Neurol 47:249–253
Kwok JB, Halliday GM, Brooks WS, Dolios G, Laudon H, Murayama O, Hallupp M, Badenhop RF, Vickers J, Wang R, Naslund J, Takashima A, Gandy SE, Schofield PR (2003) Presenilin-1 mutation L271V results in altered exon 8 splicing and Alzheimer's disease with non-cored plaques and no neuritic dystrophy. J Biol Chem 278:6748–6754
Leissring MA, Akbari Y, Fanger CM, Cahalan MD, Mattson MP, LaFerla FM (2000) Capacitative calcium entry deficits and elevated luminal calcium content in mutant presenilin-1 knockin mice. J Cell Biol 149:793–798.
Levy-Lahad E, Wasco W, Poorkaj P, Romano DM, Oshima J, Pettingell WH, Yu CE, Jondro PD, Schmidt SD, Wang K, Crowley AC, Fu YH, Guenette SY, Galas D, Nemens E, Wijsman EM, Bird TD, Schellenberg, Tanzi RE (1995a) Candidate gene for the chromosome 1 familial Alzheimer's disease locus. Science 269:973–977
Levy-Lahad E, Lahad A, Wijsman EM, Schellenberg GD (1995b) Apolipoprotein-E genotypes and age of onset in early-onset familial Alzheimer's disease. Ann Neurol 38:678–680
Lippa CF, Fujiwara H, Mann DM, Giasson B, Baba M, Schmidt ML, Nee LE, O'Connell B, Pollen DA, St George-Hyslop P, Ghetti B, Nochlin D, Bird TD, Cairns NJ, Lee VM, Iwatsubo T, Trojanowski JQ (1998) Lewy bodies contain altered alpha-synuclein in brains of many familial

Alzheimer's disease patients with mutations in presenilin and amyloid precursor protein genes. Am J Pathol 153:1365–1370

Lleo A, Castellvi M, Blesa R, Oliva R (2002) Uncommon polymorphism in the presenilin genes in human familial Alzheimer's disease: not to be mistaken with a pathogenic mutation. Neurosci Lett 318:166–168

Lleo A, Berezovska O, Growdon JH, Hyman BT (2004) Clinical, pathological, and biochemical spectrum of Alzheimer disease associated with PS-1 mutations. Am J Geriatr Psychiat 2:146–156

Lopera F, Ardilla A, Martinez A, Madrigal L, Arango-Viana JC, Lemere CA, Arango-Lasprilla JC, Hincapie L, Arcos-Burgos M, Ossa JE, Behrens IM, Norton J, Lendon C, Goate AM, Ruiz-Linares A, Rosselli M, Kosik KS (1997) Clinical features of early-onset Alzheimer disease in a large kindred with an E280A presenilin-1 mutation. JAMA 277:793–799

Mann DMA, Pickering-Brown SM, Takeuchi A, Iwatsubo T, and the Familial Alzheimer's Disease Pathology Study Group (2001) Amyloid angiopathy and variability in amyloid deposition are determined by mutation position in Presenilin-1 linked Alzheimer's Disease. Am J Pathol 158: 2165–2175

Marcon G, Giaccone G, Cupidi C, Balestrieri M, Beltrami CA, Finato N, Bergonzi P, Sorbi S, Bugiani O, Tagliavini F (2004) Neuropathological and clinical phenotype of an Italian Alzheimer family with M239V mutation of presenilin 2 gene. Neuropathol Exp Neurol 63:199–209

Moretti P, Lieberman AP, Wilde EA, Giordani BI, Kluin KJ, Koeppe RA, Minoshima S, Kuhl DE, Seltzer WK, Foster NL (2004) Novel insertional presenilin 1 mutation causing Alzheimer disease with spastic paraparesis. Neurology 62:1865–1858

Mullan M, Crawford F, Axelman K, Houlden H, Lilius L, Winblad B, Lannfelt L (1992) A pathogenic mutation for probable Alzheimer's disease in the APP gene at the N-terminus of beta-amyloid. Nature Genet 1:345–347

Murayama O, Tomita T, Nihonmatsu N, Murayama M, Sun X, Honda T, Iwatsubo T, Takashima A (1999) Enhancement of amyloid beta 42 secretion by 28 different presenilin 1 mutations of familial Alzheimer's disease. Neurosci Lett 265:61–63

Murrell J, Farlow M, Ghetti B, Benson MD (1991) A mutation in the amyloid precursor protein associated with hereditary Alzheimer's Disease. Science 254:97–99

Nilsberth C, Westlind-Danielsson A, Eckman CB, Condron MM, Axelman K, Forsell C, Stenh C, Luthman J, Teplow DB, Younkin SG, Naslund J, Lannfelt L (2001) The 'Arctic' APP mutation (E693G) causes Alzheimer's disease by enhanced Abeta protofibril formation. Nature Neurosci 4:887–893

Nochlin D, Bird TD, Nemens EJ, Ball MJ, Sumi SM (1998) Amyloid angiopathy in a Volga German family with Alzheimer's disease and a presenilin-2 mutation (N141I). Ann Neurol 43:131–135

Pastor P, Roe CM, Villegas A, Bedoya G, Chakraverty S, Garcia G, Tirado V, Norton J, Rios S, Martinez M, Kosik KS, Lopera F, Goate AM (2003) Apolipoprotein E epsilon4 modifies Alzheimer's disease onset in an E280A PS1 kindred. Ann Neurol 53:163–169

Poorkaj P, Sharma V, Anderson L, Nemens E, Alonso ME, Orr H, White J, Heston L, Bird TD, Schellenberg GD (1998) Missense mutations in the chromosome 14 familial Alzheimer's disease presenilin 1 gene. Human Mutat 11:216–221

Raux G, Gantier R, Thomas-Anterion C, Boulliat J, Verpillat P, Hannequin D, Brice A, Frebourg T, Campion D (2000) Dementia with prominent frontotemporal features associated with L113P presenilin 1 mutation. Neurology 55:1577–1578

Rossor MN, Fox NC, Beck J, Campbell TC, Collinge J (1996) Incomplete penetrance of familial Alzheimer's disease in a pedigree with a novel presenilin-1 gene mutation. Lancet 347:1560.

Scheuner D, Eckman C, Jensen M, Song X, Citron M, Suzuki N, Bird TD, Hardy J, Hutton M, Kukull W, Larson E, Levy-Lahad E, Viitanen M, Peskind E, Poorkaj P, Schellenberg G, Tanzi R, Wasco W, Lannfelt L, Selkoe D, Younkin S (1996) Secreted amyloid beta-protein similar to that in the senile plaques of Alzheimer's disease is increased in vivo by the presenilin 1 and 2 and APP mutations linked to familial Alzheimer's disease. Nature Med 8:864–870

Sherrington R, Rogaev EI, Liang Y, Rogaeva EA, Levesque G, Ikeda M, Chi H, Lin C, Li G, Holman K, Tsuda T, Mar L, Foncin JF, Bruni AC, Montesi MP, Sorbi S, Rainero I, Pinessi L, Nee L, Chumakov I, Pollen D, Brookes A, Sanseau P, Polinsky RJ, Wasko W, Da Silva HAR, Haines JL, Pericak-Vance MA, Tanzi RE, Roses AD, Fraser PE, Rommens JM, George-Hyslop PH (1995) Cloning of a gene bearing missense mutations in early-onset familial Alzheimer's disease. Nature 375:754–760

Steiner H, Romig H, Grim MG, Philipp U, Pesold B, Citron M, Baumeister R, Haass C (1999) The biological and pathological function of the presenilin-1 Deltaexon 9 mutation is independent of its defect to undergo proteolytic processing. J Biol Chem 274:7615–7618

Steiner H, Revesz T, Neumann M, Romig H, Grim MG, Pesold B, Kretzschmar HA, Hardy J, Holton JL, Baumeister R, Houlden H, Haass C (2001) A pathogenic presenilin-1 deletion causes abberrant Abeta 42 production in the absence of congophilic amyloid plaques. J Biol Chem 276:7233–7239

Tysoe C, Whittaker J, Xuereb J, Cairns NJ, Cruts M, Van Broeckhoven C, Wilcock G, Rubinsztein DC (1998) A presenilin-1 truncating mutation is present in two cases with autopsy-confirmed early-onset Alzheimer disease. Am J Human Genet 62:70–76

Vassar R, Citron M (2000) Abeta-generating enzymes: recent advances in beta- and gamma-secretase research. Neuron 27:419–422

Wakutani Y, Watanabe K, Adachi Y, Wada-Isoe K, Urakami K, Ninomiya H, Saido TC, Hashimoto T, Iwatsubo T, Nakashima K (2004) Novel amyloid precursor protein gene missense mutation (D678N) in probable familial Alzheimer's disease. J Neurol Neurosurg Psychiat 75:1039–1042

Walsh DM, Klyubin I, Fadeeva JV, Cullen WK, Anwyl R, Wolfe MS, Rowan MJ, Selkoe DJ (2002) Naturally secreted oligomers of amyloid beta protein potently inhibit hippocampal long term potentiation in vivo. Nature 416:535–539

Wen PH, Hof PR, Chen X, Gluck K, Austin G, Younkin SG, Younkin LH, DeGasperi R, Gama Sosa MA, Robakis NK, Haroutunian V, Elder GA (2004) The presenilin-1 familial Alzheimer disease mutant P117L impairs neurogenesis in the hippocampus of adult mice. Exp Neurol 188:224–237

Yokota O, Terada S, Ishizu H, Ujike H, Ishihara T, Nakashima H, Yasuda M, Kitamura Y, Ueda K, Checler F, Kuroda S (2002) NACP/alpha-synuclein, NAC, and beta-amyloid pathology of familial Alzheimer's disease with the E184D presenilin-1 mutation: a clinicopathological study of two autopsy cases. Acta Neuropathol (Berl) 104:637–648

Identification of Genes that Modify the Age of Onset in a Large Familial Alzheimer's Disease Kindred

Alison M. Goate[1]*, *Pau Pastor*[1], *Catherine M. Roe*[2], *Joanne Norton*[1], *Sumi Chakraverty*[1], *Francisco Lopera*[3], *Kenneth S. Kosik*[4]

Summary

We have identified more than 20 nuclear Alzheimer's disease (AD) families in Colombia that carry a single point mutation, E280A, in the presenilin 1 (*PSEN1*) gene. Genealogical and genetic studies have demonstrated that these families share a common founder more than 400 years ago. The mean age of onset of AD in these families is 45.2 years but the range is almost 30 years (35-62 years). The wide range in age of onset suggests that genetic and/or environmental risk factors modify the age of onset. We are using two complementary approaches to the identification of genetic modifiers: candidate gene analyses and a whole genome screen analysis.

We have genotyped polymorphisms in each of the known AD genes: *ß-amyloid protein precursor* gene (APP), *presenilin 1* (*PSEN*1) gene, *presenilin 2* (*PSEN*2) gene and *apolipoprotein E* gene (*APOE*). Several polymorphisms in these genes have previously been associated with risk for AD in some but not all studies. To determine whether these polymorphisms modify the age of onset in the Colombian kindreds, we initially used survival curve analysis. The only gene that modified age of onset with this methodology was *APOE*. The *APOE4* allele was associated with an earlier age of onset whereas the presence of the *APOE*2 allele was associated with later age of onset. This result is consistent with the observations of the effect of *APOE* alleles on sporadic AD and suggests that genes that influence the risk for late onset AD may also modify age of onset in familial early onset Alzheimer's disease (FAD).

We have also used survival analysis to examine several putative environmental risk factors. These studies demonstrated that both years of education and urban dwelling were associated with an earlier age of onset. Since these two factors were highly correlated in this study, it was not possible to determine which one was primarily the responsible risk factor.

To identify novel genetic risk factors for AD, we have used two complementary approaches: genetic analysis of a large series of late onset AD sibling pairs and a

[1] Department of Psychiatry, [2] Division of Biostatistics, and Alzheimer's Disease Research Center, Washington University School of Medicine, St. Louis, MO, [3] Grupo de Neurociencias. Facultad de Medicina, Universidad de Antioquía, Medellín, Colombia, [4] Center for Neurological Diseases, Harvard Medical School, Brigham and Women's Hospital, Boston, MA

* Department of Psychiatry, Washington University School of Medicine, Campus Box 8134, 660 South Euclid Avenue, St. Louis, MO 63110

Cummings et al.
Genotype – Protéotype – Phenotype
Relationships in Neurodegenerative Diseases

search for genes that modify age of onset in the Colombian families. Our genome screen in AD sibling pairs has provided evidence of AD susceptibility genes on chromosomes 9, 10, and 12 and suggestive evidence of an age of onset modifier gene on chromosome 12.

Introduction

Genes involved in FAD

Familial early onset AD (FAD) is a heterogeneous disorder that can be caused by mutations in at least three different genes: ß-amyloid protein precursor gene (APP) on chromosome 21, presenilin 1 (PSEN1) gene on chromosome 14, and presenilin 2 (*PSEN2*) gene on chromosome 1 (Goate et al., 1991; Levy-Lahad et al., 1995; Sherrington et al., 1995). The APP gene accounts for only a minority of cases of autosomal dominant AD whereas mutations in the *PSEN1* and *PSEN2* genes are thought to cause up to 80% of FAD cases. To date, more than 130 *PSEN1* mutations but less than 10 missense mutations in *PSEN2* have been identified. Most of the *PSEN1* mutations are missense mutations predominantly located in the highly conserved transmembrane domains. The *PSEN1* mutations are associated with an age of onset younger than 60 years and complete penetrance (Rosenberg, 2000).

The presenilin genes code for highly homologous, multi-spanning transmembrane proteins, with a high degree of conservation among species. Presenilins are located in intracellular membranes and at the cell surface (Kovacs et al., 1996) and are predicted to have eight transmembrane domains and a hydrophilic intracellular loop region. PSEN undergo a physiologic endoproteolytic cleavage to generate stable N- and C- fragments (Thinakaran et al., 1997). There is growing evidence that PSEN form the catalytic center of γ-secretase, a multi-protein aspartyl protease. In vivo and in vitro experiments have demonstrated that mutant PSEN increase Aß42 levels (Borchelt et al., 1996; Wolfe et al., 1999). Functional studies suggest that PSEN mutations lead to a gain of PSEN protein function with respect to Aß generation, which is consistent with the inheritance pattern of early-onset FAD cases. Some FAD mutations in PSEN exhibit a partial loss of function with respect to γ-secretase cleavage of other substrates (Walker et al., 2005).

APOE polymorphisms

Apolipoprotein E *(APOE)* exists as three isoforms in all populations, although the relative frequency of the isoforms can vary (Zannis et al., 1981). *APOE*3 is the major isoform in all populations. *APOE*2 and *APOE*4 vary from *APOE*3 at residues 112 and 158. Approximately 50% of AD cases carry one or more *APOE4* alleles (Henderson et al., 1995; Martins et al., 1995; Kukull et al., 1996). *APOE4* shows a dose-dependent increase in risk for sporadic AD: heterozygotes have a three-fold increased risk for disease and homozygotes have an eight-fold increased risk for disease in Caucasians (Henderson et al., 1995; Martins et al., 1995; Kukull et al.,

1996; Corbo and Scacchi, 1999). Most individuals who are homozygous for the *APOE4* allele develop AD by age 80 (Corder et al., 1993). *APOE4* alleles are associated with an earlier age of onset, with each allele causing a decrease in onset of approximately five years. In contrast, the *APOE2* allele decreases risk for AD and increases age of onset of disease (Corbo and Scacchi, 1999). The association between AD and *APOE4* has been confirmed in most populations worldwide (Tang et al., 1998; Graff-Radford et al., 2002). There are several in vitro and in vivo studies that have tried to reveal the role of *APOE* in the pathogenesis of AD. In vitro, APOE binds to synthetic Aß peptide (the primary constituent of the senile plaque) with high avidity (Strittmatter et al., 1993). Transgenic experiments have demonstrated that *APOE*4 increases risk for disease by promoting Aß deposition (Mahley, 1988). APOE4 most likely influences fibril formation and/or clearance of Aß, thus accelerating Aß deposition (Holtzman et al., 2000). The *APOE4* allele is strongly associated with an increased number of neuritic plaques and cerebral amyloid angiopathy in AD (Olichney et al., 1996).

Genetic modifiers of the phenotypic presentation of AD

In the last decade, several groups have attempted to dissect the clinical phenotype to try to identify a more genetically homogeneous subgroup of families and increase the statistical power to detect linkage. Among the phenotypic traits of AD, the most reliable seems to be the age of onset of disease. The identification of genetic factors that can modify the onset could lead us to new therapeutic strategies to delay the onset of the disease.

APOE and other modifiers of the age of onset in FAD

Age of onset is clearly an important covariate for the identification of genes for AD. Segregation analysis of large pedigrees with AD suggests that multiple loci associated with age of onset exist (Daw et al., 1999). In this study, when age of onset was examined as a quantitative trait, it was estimated that up to four additional major genes as well as several minor AD genes remained to be identified (Daw et al., 1999). The presence of *APOE4* alleles is correlated with earlier age of onset of AD (Poirier et al., 1993; Chartier-Harlin et al., 1994; Tsai et al., 1994; Meyer et al., 1998). Daw and colleagues analyzed 75 families with late-onset AD and found a dose effect for the *APOE*4 allele and a protective effect for the *APOE*2 allele. However, the estimate of the contribution of *APOE* to the total variation in onset of AD was between 7 and 9% (Daw et al, 2000).The role of *APOE4* in modifying age of onset has also been examined in AD families with known causative mutations. Early studies reported an association between the presence of the *APOE4* allele and age of onset in small kindreds carrying APP mutations, but not in those carrying PSEN mutations (Haan et al., 1994; Van Broeckhoven et al., 1994; Sorbi et al., 1995; Lendon et al., 1997).

In 1987, a Colombian family with early-onset autosomal dominant AD was described (Cornejo et al., 1987). Screening of the *PSEN1* gene revealed a missense

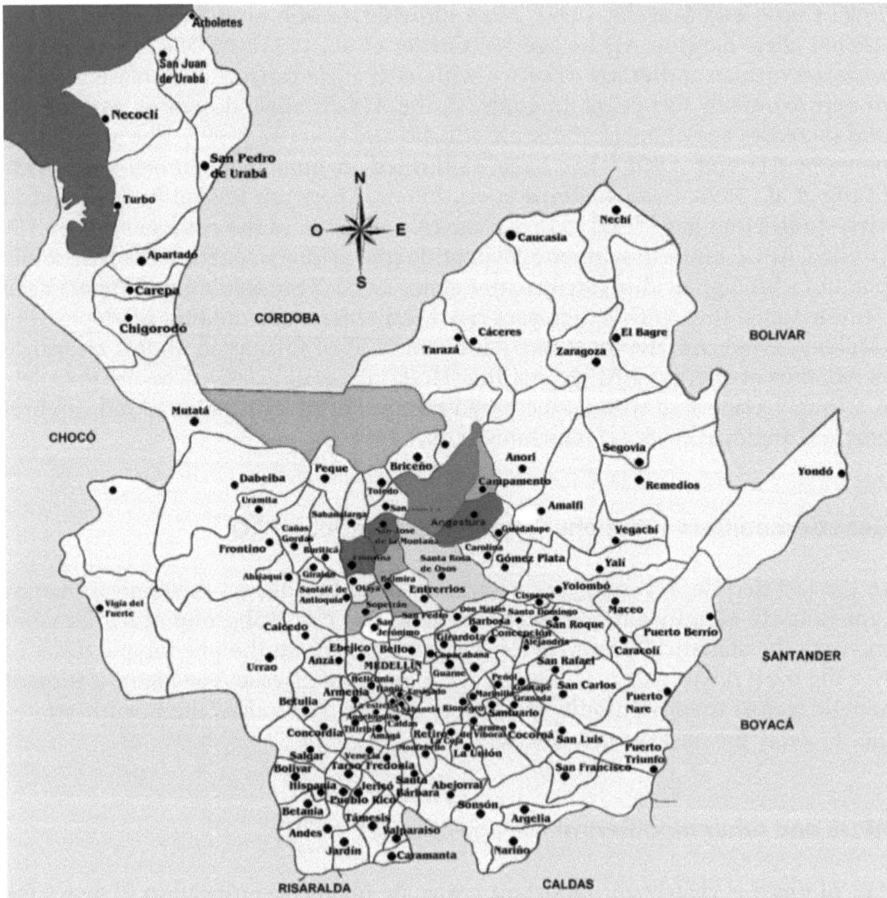

Fig. 1. Map of Colombia showing the regions in which the E280A mutation families are located.

mutation at codon 280 of *PSEN1* (Lendon et al., 1997). In subsequent years, 24 more families with the E280A *PSEN1* mutation were identified from several towns and villages in the province of Antioquia (Fig. 1; Lopera et al., 1994). Genetic and genealogical studies have demonstrated that each of these families has a common founder dating back to the mid-eighteenth century (Fig. 2). This large family with similar environmental exposures and homogeneous disease etiology is a unique resource for the study of modifier genes. We first analyzed the influence of certain *APOE* isoforms in the onset of the AD symptoms (Pastor et al. 2003). A total of 109 individuals carrying a copy of the E280A *PSEN1* mutation made up the sample. Among these 109 individuals, 57 remained asymptomatic throughout the study period, and 52 individuals developed or already had AD symptoms. Age of onset ranged from 35-62 years, with a mean of 45.2 years (SD: 5.7), and 75% of AD patients had an age of onset between 39 and 51 years. For patients without AD, age

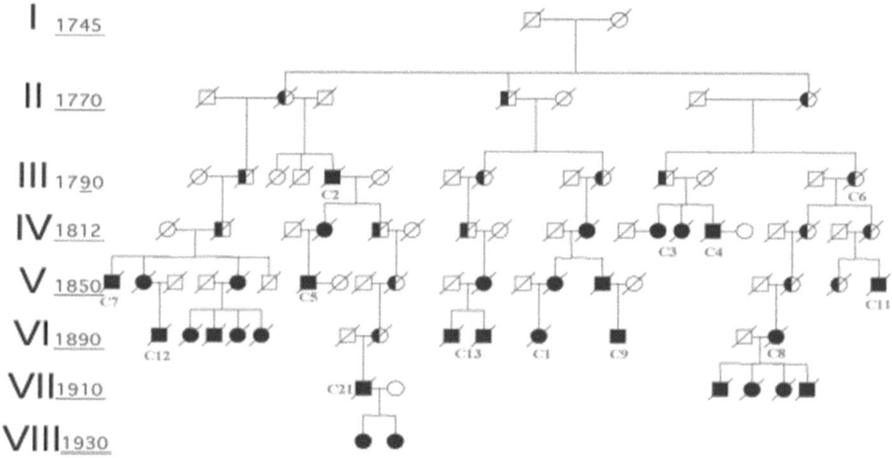

Fig. 2. Pedigree showing how 13 of the families are linked to a common founder. Each family is numbered (C1....). Roman numerals indicate the generations in the pedigree plus the approximate year of birth in those generations.

at last assessment ranged from 24 to 65 years, with a mean of 40.5 years (SD: 8.5). *APOE* polymorphism analysis showed that only one (0.9%) of the 109 carriers of the E280A mutation was homozygous for the *APOE4* allele. Twenty-eight (25.7%) were heterozygous (N = 25 for *E4/E3* and N = 3 for *E4/ E2*) and 80 (73.4%) had no *E4* alleles (N = 72 for *E3/E3* and N = 8 for *E3/E2*).

Kaplan-Meier Product Limit and Cox Proportional Hazard Models were used in the statistical analyses of age onset in this set of E280A carriers, after subjects were dichotomized into two groups based on the presence or absence of the *APOE4* allele. The results showed that *APOE4* allele carriers were more likely to develop AD at an earlier age than subjects without the *E4* allele (hazard ratio [HR]= 2.07, 95% CI = 1.07 – 3.99, p = .030; Fig. 3A). The survival curve of time to AD onset of subjects with *E3/E2* and *E4/E2* genotypes (N = 11) was compared to the survival function of subjects with *E4/E3* and *E4/E4* genotypes (N = 26), and with all subjects without the *APOE2* allele *(N=98)* using Kaplan-Meier Product-Limit. Although the survival curves suggest a modest protective influence of the *APOE2* allele (Fig. 3B), the difference between the groups did not reach statistical significance (p = 0.106 and p = 0.15, respectively).

We also used survival analysis to evaluate the impact of several demographic factors on age of onset. Previous studies of late onset AD have suggested that a low level of education is associated with increased risk for AD (Lindsay et al.,2002; Stern et al. 1994), whereas other studies have suggested that late diagnosis of dementia is associated with lower levels of formal education, rural residence and lower occupational role (Antonelli et al. 1992). To investigate whether educational level was associated with age of AD onset within this kindred, subjects were stratified into low (0 to 3 years, N = 52) and high (more than 3 years, N = 48) education groups. The mean educational level of subjects was low (4.6 years, SD: 3.5 years).

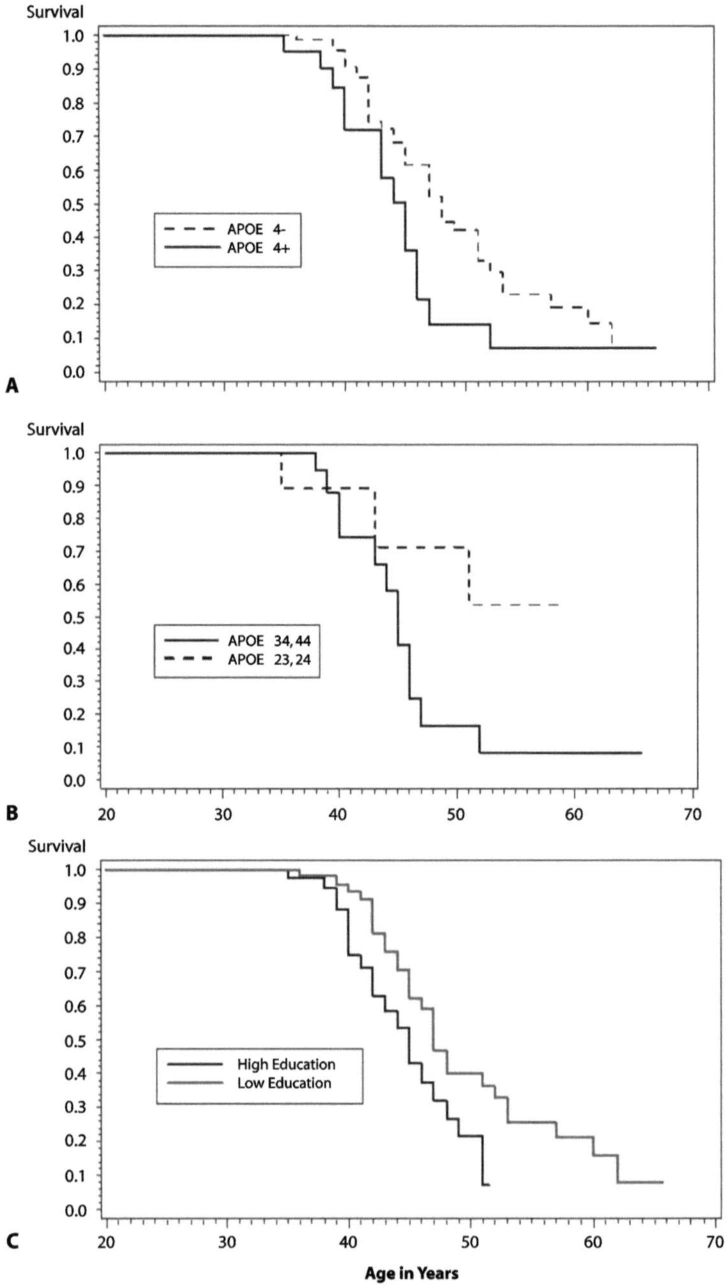

Fig. 3. Survival curves of age of AD onset for subjects carrying the E280A *PSEN1* mutation. **A** With (*APOE 4+*) and without the *APOE4* allele(*APOE 4-*). **B** With the *E3/E2* or *E4/E2* genotypes versus with the *E4/E3* or *E4/E4* genotypes. **C** With high and low levels of education (Pastor et al., 2003).

Kaplan-Meier analyses indicated that subjects in the low education group were more likely to develop AD at a later age than were subjects in the high education group (log-rank test, $p = 0.017$; Fig. 3C) and that subjects residing in a rural area were more likely to develop AD at a later age than were those who lived in an urban setting (log-rank test, $p = 0.007$; Fig. 3C). Educational level was related to residential area (chi-square test, $p < .001$), such that 68.6% of individuals in the low education group lived in a rural area, compared with 35.4% of those in the high education group. A regression model testing *APOE4* status, education group, and the interaction term indicated that *APOE4* status and education did not have an interactive effect ($p = 0.278$) on age of onset. This finding demonstrates that the APOE isoform and environmental factors independently influence age of onset in a kindred with an FAD mutation.

Interestingly, Wijsman and colleagues (2004) have recently found a significant but small role of the APOE isoforms in modifying the age of onset in several kindreds of Volga German ancestry with a *PSEN2* mutation. These authors also estimated an 83% posterior probability of at least one age of onset modifier locus in addition to *APOE*.

Other candidate genes

Several recent studies have shown that certain variants in the regulatory region of the *PSEN1* and *PSEN2* genes can be associated with risk for AD (Theuns et al., 2000; Riazanskaia et al., 2002). We used the same statistical approach as for the APOE analysis to investigate the possible role of those single nucleotide polymorphisms (SNPs) in age of onset in the Colombian AD kindreds. Neither the *PSEN1* -48C/T nor the *PSEN2* 24914delA variants showed a significant effect on the age of onset. Similarly, a polymorphism within the APP gene failed to show any effect on the age of onset. Since each of these studies only analyzed one SNP in each gene, these results do not preclude the possibility that other polymorphisms within these genes may influence age of onset. To test this hypothesis we are carrying out a rigorous analysis of multiple SNPs in each gene in late onset AD cases. If we are able to establish evidence of association with specific SNPs or haplotypes in these AD cases, we will re-evaluate these genes in the Colombian kindreds.

Effect of other candidate chromosomal regions on FAD onset

We have also used several approaches to test for genes that modify age of onset in late onset AD. One approach is to use age of onset as a quantitative trait locus (QTL) in either parametric or nonparametric analyses. A second approach is to include age of onset as a covariate. Using age of onset as a QTL in our late onset AD sibling pair data, we observed evidence of linkage on chromosomes 7, 12 and 19. The peak on chromosome 19 localizes close to *APOE* and the locus on chromosome 12 is in the same position as the peak observed in our original genome screen in late onset AD (LOAD;(Myers et al. 2002). When mean age of onset was included as a covariate, the strongest evidence for linkage was observed on chromosome 21 [Lod score (LOD) increase = 2.01], whereas analyses using age of onset difference as a covariate resulted in suggestive evidence of linkage on the short arm of chromosome 12 (LOD increase =1.38). Interestingly, both of these studies support the previous evidence for an AD susceptibility gene on the short arm of chromosome 12. Recently, we have performed a large-scale, SNP-based association study in this region of chromosome 12 in three large case control series (Li et al. 2004). Several SNPs in the glyceraldehyde-3-phosphate dehydrogenase (*GAPD*) gene showed evidence of association in two of the three case control series, raising the possibility that *GAPD* genes may influence risk for AD.

Other studies have also used age of onset as a covariate in a genome-wide screen (Pericak-Vance et al., 2000; Scott et al., 2003) and observed significant non-parametric multipoint LOD scores for several different intervals of age of onset. They observed a LOD score of 3.2 at D2S2944 on chromosome 2q34 in families with an age of onset between 50 and 60 years and a LOD score of 4.6 at D9S741 in the chromosome 9p region previously linked to AD in families with an age of onset between 60 and 75 years. A LOD score of 2.8 was detected at D15S1507 with age of onset >/=79 years, and a peak LOD score of 3.1 was obtained at D15S153 (62 cM) in families with mean age of onset >80 years. Another study, using a combined data set of FAD and familial Parkinson disease families, reported evidence for a locus on chromosome 10q controlling the age of onset of both disorders (Li et al., 2002).

Conclusions

FAD is a genetically heterogeneous disorder. Mendelian forms of the disease show genetic heterogeneity with at least three genes that act through a common biochemical pathway to cause disease. To date, the *APOE4* allele is the only established modifier of genetic risk in late-onset AD. Survival analyses and segregation analyses in families with *PSEN1* and *PSEN2* mutations demonstrate that *APOE* alleles also influence the age of onset in FAD. In a *PSEN1* mutation family, we also demonstrate evidence of the role of environmental factors in the age of onset. Segregation analyses in LOAD and *PSEN2* families suggest that there are additional genes that modify age of onset. Studies from our own lab and from others have implicated several other chromosomal regions but no other specific genes have yet been identified.

Acknowledgments

The authors thank all the family members of the Colombian kindred for their support of this project. This work was supported by COLCIENCIAS grant, 115-04-007-99 and NIA grants, AG16208 and P50 AG05681.

References

Antonelli IR, Marra C, Gemma A, Capparella O, Carbonin PU (1992) Unrecognized dementia: sociodemographic correlates. Aging Clin Exp Res 4: 327-332.

Borchelt DR, Thinakaran G, Eckman CB, Lee MK, Davenport F, Ratovitsky T, Prada CM, Kim G, Seekins S, Yager D, Slunt HH, Wang R, Seeger M, Levey AI, Gandy SE, Copeland NG, Jenkins NA, Price DL, Younkin SG, Sisodia SS (1996) Familial Alzheimer's disease-linked presenilin 1 variants elevate Abeta1-42/1-40 ratio in vitro and in vivo. Neuron 17:1005-1013.

Chartier-Harlin MC, Parfitt M, Legrain S, Perez-Tur J, Brousseau T, Evans A, Berr C, Vidal O, Roques P, Gourlet V, Fruchart J-C, Delacourte A, Rossor M, Amouyel P (1994) Apolipoprotein E, epsilon 4 allele as a major risk factor for sporadic early and late-onset forms of Alzheimer's disease: analysis of the 19q13.2 chromosomal region. Hum Mol Genet 3:569-574.

Corbo RM, Scacchi R (1999) Apolipoprotein E (APOE) allele distribution in the world. Is APOE*4 a 'thrifty' allele? Ann Hum Genet 63 (Pt 4):301-310.

Corder EH, Saunders AM, Strittmatter WJ, Schmechel DE, Gaskell PC, Small GW, Roses AD, Haines JL, Pericak-Vance MA (1993) Gene dose of apolipoprotein E type 4 allele and the risk of Alzheimer's disease in late onset families. Science 261:921-923.

Cornejo W, Lopera F, Uribe CS, Salinas M (1987) Descripci6n de una familia con demencia presenil tipo Alzheimer. Acta Med Colombiana 12:55-61.

Daw EW, Heath SC, Wijsman EM (1999) Multipoint oligogenic analysis of age-at-onset data with applications to Alzheimer disease pedigrees. Am J Hum Genet 64:839-851.

Goate A, Chartier-Harlin MC, Mullan M, Brown J, Crawford F, Fidani L, Giuffra L, Haynes A, Irving N, James L, Mant R, Newton P, Rook, K, Roques P, Talbot C, Pericak-Vance M, Roses A, Williamson R, Rossor M, Owen M, Hardy J (1991) Segregation of a missense mutation in the amyloid precursor protein gene with familial Alzheimer's disease. Nature 349:704-706.

Graff-Radford NR, Green RC, Go RC, Hutton ML, Edeki T, Bachman D, Adamson JL, Griffith P, Willis FB, Williams M, Hipps Y, Haines JL, Cupples LA, Farrer LA (2002) Association between apolipoprotein E genotype and Alzheimer disease in African American subjects. Arch Neurol 59:594-600.

Haan J, Van Broeckhoven C, van Duijn CM, Voorhoeve E, van Harskamp F, van Swieten JC, Maat-Schieman ML, Roos RA, Bakker E (1994) The apolipoprotein E epsilon 4 allele does not influence the clinical expression of the amyloid precursor protein gene codon 693 or 692 mutations. Ann Neurol 36:434-437.

Henderson AS, Easteal S, Jorm AF, Mackinnon AJ, Korten AE, Christensen H, Croft L, Jacomb PA (1995) Apolipoprotein E allele epsilon 4, dementia, and cognitive decline in a population sample. Lancet 346:1387-1390.

Holtzman DM, Fagan AM, Mackey B, Tenkova T, Sartorius L, Paul SM, Bales K, Ashe KH, Irizarry MC, Hyman BT (2000) Apolipoprotein E facilitates neuritic and cerebrovascular plaque formation in an Alzheimer's disease model. Ann Neurol 47:739-747.

Kovacs DM, Fausett HJ, Page KJ, Kim TW, Moir RD, Merriam DE, Hollister RD, Hallmark OG, Mancini R, Felsenstein KM, Hyman BT, Tanzi RE, Wasco W (1996) Alzheimer-associated presenilins 1 and 2: neuronal expression in brain and localization to intracellular membranes in mammalian cells. Nat Med 2:224-229.

Kukull WA, Schellenberg GD, Bowen JD, McCormick WC, Yu CE, Teri L, Thompson JD, O'Meara ES, Larson EB (1996) Apolipoprotein E in Alzheimer's disease risk and case detection: a case-control study. J Clin Epidemiol 49:1143-1148.

Lendon CL, Martinez A, Behrens IM, Kosik KS, Madrigal L, Norton J, Neuman R, Myers A, Busfield F, Wragg M, Arcos M, Arango Viana JC, Ossa J, Ruiz A, Goate AM, Lopera F (1997) E280A PS-1 mutation causes Alzheimer's disease but age of onset is not modified by ApoE alleles. Hum Mutat 10:186-195.

Levy-Lahad E, Wasco W, Poorkaj P, Romano DM, Oshima J, Pettingell WH, Yu CE, Jondro PD, Schmidt SD, Wang K, Crowley AC; Fu YH, Guenette SY, Galas D, Nemens E, Wijsman EM,

Bird TD, Schellenberg GD, Tanzi RE (1995) Candidate gene for the chromosome 1 familial Alzheimer's disease locus. Science 269:973-977.

Li YJ, Scott WK, Hedges DJ, Zhang F, Gaskell PC, Nance MA, Watts RL, Hubble JP, Koller WC, Pahwa R, Stern MB, Hiner BC, Jankovic J, Allen FA, Jr., Goetz CG, Mastaglia F, Stajich JM, Gibson RA, Middleton LT, Saunders AM, Scott BL, Small GW, Nicodemus KK, Reed AD, Schmechel DE, Welsh-Bohmer KA, Conneally PM, Roses AD, Gilbert JR, Vance JM, Haines JL, Pericak-Vance MA (2002) Age at onset in two common neurodegenerative diseases is genetically controlled. Am J Hum Genet 70:985-993.

Li Y, Nowotny P, Holmans P, Smemo S, Kauwe J, Tacey K, Doil L, van Luchen R, Garcia V, Rowland C, Schrodi S, Leong D, Gogic G, Chan J, Cravchik A, Ross D, Lau K, Kwok S, Chang S-Y, Catanes J, Sninsky J, White T, Hardy J, Powell J, Lovestone S, Thal L, Owen M, Williams J, Goate A, Grupe A (2004) Association of late onset Alzheimer's disease with genetic variation in multiple members of the glyceraldehyde-3-phosphate dehydrogenase gene family. Proc. Natn. Acad. Sci. USA 101:15688-15693

Lopera F, Arcos-Burgos M, Madrigal L, Kosik KS, Cornejo W, Ossa J (1994) Demencia tipo Alzheimer con Agregación familiar en Antioquia, Colombia. Acta Neurológica Colombiana 10:173-187.

Mahley RW (1988) Apolipoprotein E: cholesterol transport protein with expanding role in cell biology. Science 240:622-630.

Martins RN, Clarnette R, Fisher C, Broe GA, Brooks WS, Montgomery P, Gandy SE (1995) ApoE genotypes in Australia: roles in early and late onset Alzheimer's disease and Down's syndrome. Neuroreport 6:1513-1516.

Meyer MR, Tschanz JT, Norton MC, Welsh-Bohmer KA, Steffens DC, Wyse BW, Breitner JC (1998) APOE genotype predicts when–not whether–one is predisposed to develop Alzheimer disease. Nat Genet 19:321-322.

Olichney JM, Hansen LA, Galasko D, Saitoh T, Hofstetter CR, Katzman R, Thal LJ (1996) The apolipoprotein E epsilon 4 allele is associated with increased neuritic plaques and cerebral amyloid angiopathy in Alzheimer's disease and Lewy body variant. Neurology 47:190-196.

Pastor P, Roe CM, Villegas A, Bedoya G, Chakraverty S, Garcia G, Tirado V, Norton J, Rios S, Martinez M, Kosik KS, Lopera F, Goate AM (2003) Apolipoprotein Eepsilon4 modifies Alzheimer's disease onset in an E280A PS1 kindred. Ann Neurol 54:163-169.

Pericak-Vance MA, Grubber J, Bailey LR, Hedges D, West S, Santoro L, Kemmerer B, Hall JL, Saunders AM, Roses AD, Small GW, Scott WK, Conneally PM, Vance JM, Haines JL (2000) Identification of novel genes in late-onset Alzheimer's disease. Exp Gerontol 35:1343-1352.

Poirier J, Davignon J, Bouthillier D, Kogan S, Bertrand P, Gauthier S (1993) Apolipoprotein E polymorphism and Alzheimer's disease. Lancet 342:697-699.

Riazanskaia N, Lukiw WJ, Grigorenko A, Korovaitseva G, Dvoryanchikov G, Moliaka Y, Nicolaou M, Farrer L, Bazan NG, Rogaev E (2002) Regulatory region variability in the human presenilin-2 (PSEN2) gene: potential contribution to the gene activity and risk for AD. Mol Psychiatry 7:891-898.

Rosenberg RN (2000) The molecular and genetic basis of AD: the end of the beginning: the 2000 Wartenberg lecture. Neurology 54:2045-2054.

Scott WK, Hauser ER, Schmechel DE, Welsh-Bohmer KA, Small GW, Roses AD, Saunders AM, Gilbert JR, Vance JM, Haines JL, Pericak-Vance MA (2003) Ordered-subsets linkage analysis detects novel Alzheimer disease Loci on chromosomes 2q34 and 15q22. Am J Hum Genet 73:1041-1051.

Sherrington R, Rogaev EI, Liang Y, Rogaeva EA, Levesque G, Ikeda M, Chi H, Lin C, Li G, Holman K, Tsuda T, Mar L, Foncin J-F, Bruni AC, Montesi MP, Sorbi S, Rainero I, Pinessi L, Nee L, Chumakov I, Pollen D, Brookes A, Sanseau P, Polinsky RJ, Wasco W, Da Silva HAR, Haines JL, Pericak-Vance MA, Tanzi RE, Roses AD, Fraser PE, Rommens JM, St George-Hyslop PH (1995) Cloning of a gene bearing missense mutations in early-onset familial Alzheimer's disease. Nature 375:754-760.

Sorbi S, Nacmias B, Forleo P, Piacentini S, Latorraca S, Amaducci L (1995) Epistatic effect of APP717 mutation and apolipoprotein E genotype in familial Alzheimer's disease. Ann Neurol 38:124-127.

Strittmatter WJ, Saunders AM, Schmechel D, Pericak-Vance M, Enghild J, Salvesen GS, Roses AD (1993) Apolipoprotein E: high-avidity binding to beta-amyloid and increased frequency of type 4 allele in late-onset familial Alzheimer disease. Proc Natl Acad Sci U S A 90:1977-1981.

Tang MX, Stern Y, Marder K, Bell K, Gurland B, Lantigua R, Andrews H, Feng L, Tycko B, Mayeux R (1998) The APOE-epsilon4 allele and the risk of Alzheimer disease among African Americans, whites, and Hispanics. Jama 279:751-755.

Theuns J, Del-Favero J, Dermaut B, van Duijn CM, Backhovens H, Van den Broeck MV, Serneels S, Corsmit E, Van Broeckhoven CV, Cruts M (2000) Genetic variability in the regulatory region of presenilin 1 associated with risk for Alzheimer's disease and variable expression. Hum Mol Genet 9:325-331.

Thinakaran G, Harris CL, Ratovitski T, Davenport F, Slunt HH, Price DL, Borchelt DR, Sisodia SS (1997) Evidence that levels of presenilins (PS1 and PS2) are coordinately regulated by competition for limiting cellular factors. J Biol Chem 272:28415-28422.

Tsai MS, Tangalos EG, Petersen RC, Smith GE, Schaid DJ, Kokmen E, Ivnik RJ, Thibodeau SN (1994) Apolipoprotein E: risk factor for Alzheimer disease. Am J Hum Genet 54:643-649.

Van Broeckhoven C, Backhovens H, Cruts M, Martin JJ, Crook R, Houlden H, Hardy J (1994) APOE genotype does not modulate age of onset in families with chromosome 14 encoded Alzheimer's disease. Neurosci Lett 169:179-180.

Walker ES, Martinez M, Brunkan AL, Goate A (2005) Presenilin 2 familial Alzheimer's disease mutations result in partial loss of functin and dramatic changes in Abeta 42/40 ratios. J Neurochem 92:294-301.

Wolfe MS, Xia W, Ostaszewski BL, Diehl TS, Kimberly WT, Selkoe DJ (1999) Two transmembrane aspartates in presenilin-1 required for presenilin endoproteolysis and gamma-secretase activity. Nature 398:513-517.

Zannis VI, Just PW, Breslow JL (1981) Human apolipoprotein E isoprotein subclasses are genetically determined. Am J Hum Genet 33:11-24.

Variable Phenotype of Alzheimer's Disease with Spastic Paraparesis

Helena Karlstrom[1], *William S Brooks*[2,4], *John BJ Kwok*[1], *Jillian J Kril*[3], *Glenda M Halliday*[2,4], *and Peter R Schofield*[1,2,4]

Summary

Pedigrees with familial Alzheimer's disease (FAD), caused by mutations in either the amyloid precursor protein (APP) or the presenilin 1 (PS1) or presenilin 2 (PS2) genes, show considerable phenotypic variability. Monogenic diseases typically exhibit variations in biological features, such as age of onset, severity, and multiple clinical and cellular phenotypes. This variation can be due to specific alleles of the disease gene, environmental effects, or modifier genes.

Spastic paraparesis (SP), or progressive spasticity of the lower limbs, is frequently hereditary, with over 20 loci being identified for uncomplicated (paraparesis alone) and complicated (paraparesis and other neurological features) disease subtypes. Moreover, over 10 different genes have been identified with mutations that lead to SP. While dementia is a common feature of complicated SP, a reciprocal observation has also been made since the earliest clinical reports of FAD: namely, that a number of AD families have been reported in which some individuals have SP. In 1997, the key observation was made that PS1 mutations were associated with the presence of SP, suggesting that there was a complex relationship between SP and AD. In addition, in 1998, it was also shown that PS1 AD/SP pedigrees frequently have variant, large, non-cored plaques without neuritic dystrophy, named cotton wool plaques (CWP). The PS1 mutations associated with CWP secrete unusually high levels of the amyloid β 42 peptide, suggesting a molecular basis for the formation of this distinctive plaque type.

The SP phenotype in PS1 pedigrees appears to be associated in some cases with a delayed onset of dementia, compared with affected individuals who present with dementia only. Some individuals who present with SP have remained dementia-free for up to 10 years. Variations seen in neuropathology and neurological symptoms in PS1 FAD suggest that modifier genes may underlie this phenotypic heterogeneity. As PS1 mutations are almost always associated with a particularly aggressive form of presenile dementia, these findings suggest the existence of a protective factor in some individuals with SP.

[1] Garvan Institute of Medical Research, Sydney, Australia
[2] Prince of Wales Medical Research Institute, Sydney, Australia
[3] Centre for Education and Research on Ageing, University of Sydney, Sydney, Australia
[4] University of New South Wales, Sydney, Australia.

Cummings et al.
Genotype – Protéotype – Phenotype
Relationships in Neurodegenerative Diseases

Introduction

Alzheimer's disease (AD) is one of the most common neurodegenerative diseases, with a prevalence of 5–10% in individuals over 65 years of age. Approximately 70% of all cases of dementia are due to AD (Hardy 1997) AD is characterised by memory loss associated with neuronal degeneration, amyloid plaques and cholinergic deficiency in many areas of grey matter, with the death of neurons accompanied by formation of abnormal cytoskeletal structures (neurofibrillary tangles). The duration of AD varies widely between 5 and 15 years, and the disease represents a large medical, social and economic problem.

The Bavarian psychiatrist, Alois Alzheimer, described the first case of AD in 1907. Post-mortem examination of the brain revealed both senile plaques and neurofibrillary tangles (NFT), the classical hallmarks of AD, in the neocortex and hippocampus (Alzheimer 1907). Senile plaques consist of extracellular deposits of the beta amyloid (Aβ) peptide and can be present either with or without neuritic infiltrate (neuritic or diffuse plaques; Selkoe 1994) NFT consist of intracellular bundles of straight and paired helical filaments of an abnormally hyperphosphorylated form of the microtubule-associated protein, tau. NFT are believed to be a general sign of neurodegeneration, since they are evident in other neurological diseases such as progressive supranuclear palsy and frontotemporal dementia (Grundke-Iqbal et al. 1986).

The majority of AD cases present as sporadic, late-onset forms, where no familial link has been established. However, approximately 5% of all AD cases occur as an autosomal dominant familial form, which is usually characterised by an early age at onset of disease (<65 years of age). Many pedigrees with familial Alzheimer's disease (FAD) have been shown to carry mutations in one of the APP, PS1 or PS2 genes (Hardy 1997; Price and Sisodia 1998). The PS1 and PS2 genes were identified by positional cloning in 1995 and have been shown to be involved in the proteolysis of the APP protein, generating the Aβ peptide (De Strooper et al. 1999; Herreman 2000; Levy-Lahad et al. 1995a b; Rogaev et al. 1995). To date, 139 PS1 mutations, 10 PS2 mutations and 16 APP mutations have been identified that result in FAD (see AD mutation databases http://www.alzforum.org or http://www.molgen.ua.ac.be/ADMutations).

Monogenic diseases typically exhibit variations in biological features, such as age of onset, severity, and multiple clinical and cellular phenotypes. These variations can be due to specific alleles of the disease gene, environmental effects, or modifier genes. Genetic factors that modify the age of onset of AD have been proposed; the Apolipoprotein E ε4 allele results in an earlier age of onset in FAD pedigrees with APP mutations (Chartier-Harlin et al. 1994). With the increasing number of identified mutations, especially PS1 mutations, it has become apparent that there are a number of associated clinical phenotypes, all characterised by cognitive dysfunction but with various other neurological symptoms and variant pathology.

Spastic paraparesis

Spastic paraparesis (SP) is frequently hereditary and is characterised by insidiously progressive lower extremity weakness and spasticity. SP is classified as either uncomplicated or complicated ("complex"). Uncomplicated SP occurs when the neurologic impairment is limited to progressive lower extremity spastic weakness, hypertonic urinary bladder disturbance, and mild diminution of lower extremity vibration sensation and, occasionally, of joint position sensation (Harding 1983). Uncomplicated SP can begin at any age from early childhood through to late adulthood. Affected individuals experience progressive difficulty walking and often require canes, walkers, or wheelchairs. The patients retain normal strength and dexterity of the upper extremities and have no involvement of speech, chewing, or swallowing. Though typically disabling, uncomplicated SP does not shorten life span.

If the impairments present in uncomplicated SP are accompanied by other system involvement or other neurological findings, such as seizures, dementia, amyotrophy, extrapyramidal disturbances or peripheral neuropathy, the SP is classified as complicated. Neurologic examination of SP demonstrates corticospinal tract deficits affecting both lower extremities, often accompanied by mildly impaired vibration sensation in the distal lower extremities and hypertonic urinary bladder. MRIs of the brain and spinal cord are usually normal in uncomplicated SP; however, some patients have atrophy of the thoracic spinal cord (Hedera et al. 1999). The primary pathology of uncomplicated SP is axonal degeneration at the distal corticospinal tracts (Behan and Maia 1974; Harding 1993; Schwarz and Liu 1956). Family history is consistent with autosomal dominant, autosomal recessive, or X-linked recessive inheritance. Over 20 genetic loci have been identified for SP with mutations identified in genes that give rise to complicated and uncomplicated SP that is inherited in dominant, recessive and X-linked forms (Table 1; reviewed in Fink 2003). Of particular interest are the SPG7, SPG20, ARSACS and SPG15 loci, which are recessive forms of complicated SP; mutations have been identified for the first three in the paraplegin, spartin and sacsin genes, respectively (Casari et al. 1998; Engert et al. 1999; Patel et al. 2002). No gene has been identified at the SPG15 locus, but families linked to this locus present with dementia as the principal additional clinical feature (Hughes et al. 2001). Simpson et al. (2003) reported a base-pair insertion in the acid-cluster protein (ACP33) gene in an Old Order Amish family with an autosomal recessive, complicated form of hereditary SP with dementia called Mast Syndrome. The frameshift results in premature termination of the ACP33 gene, which is designated "maspardin" (Simpson et al. 2003).

Spastic paraparesis and Presenilin 1 mutations

From the very earliest reports of AD, there have been patients studied who also have SP. The first such report was by Barrett (1913), who described a patient with AD in her 30s with unusual neurological disturbances that included SP. Later, van Bogaert et al. (1940) reported a Belgian pedigree (family DC or CO) with FAD and

Table 1. Genetic loci and genes that cause spastic paraparesis (SP).

Classification	Inheritance	Locus	Location	Gene
Complicated SP	Dominant	SPG9	10q23.3-q24.1	Unknown
	Dominant	SPG17	11q12-q14	Unknown
	Dominant	SAX1	12p13	Unknown
	Recessive	SPG7	16q24.3	Paraplegin
	Recessive	SPG14	3q27-q28	Unknown
	Recessive	SPG15	14q22-q24	Unknown
	Recessive	SPG20	13q12.3	Spartin
	Recessive	Mast Syndrome	15q21-q22	Maspardin (ACP33)
	Recessive	ARSACS	13q12	Sacsin
	X-linked	SPG1	Xq28	L1CAM
	X-linked	SPG2	Xq22	PLP1
	X-linked	SPG16	Xq11.2-23	Unknown
Uncomplicated SP	Dominant	SPG3A	14q11-q21	Atlastin
	Dominant	SPG4	2p22-p21	Spastin
	Dominant	SPG6	15q11.1	NIPA1
	Dominant	SPG8	8q23-q24	Unknown
	Dominant	SPG10	12q13	KIF5A
	Dominant	SPG12	19q13	Unknown
	Dominant	SPG13	2q24	Hsp60
	Dominant	SPG19	9q33-q34	Unknown
	Recessive	SPG5A	8p12-q13	Unknown
	Recessive	SPG11	15q13-q15	Unknown
	Recessive	SPG21	13q14	Unknown

SP. In three of five affected individuals, cognitive problems preceded SP, but in the two subjects whose initial problem was gait difficulty, cognitive decline was not noted for two years and 12 years, respectively. In addition to the typical plaques and NFT of AD, both Barrett (1913) and van Bogaert et al. (1940) documented generalised atrophy of the brain together with degeneration of the corticospinal tract (Barrett 1913; van Bogaert et al. 1940).

In 1997, we reported the first definitive association of SP with PS1 mutations, thereby defining a potential genetic etiology to this neurological variant of FAD (Kwok et al. 1997). Each of the pedigrees with AD and SP that also carries a PS1 mutation is described in further detail in Table 2. A significant number of AD/SP pedigrees have PS1 exon 9 deletions, although a number of other missense muta-

Table 2. Alzheimer's disease (AD) pedigrees carrying Presenilin 1 mutations that have spastic paraparesis (SP) and cotton wool plaques (CWP).

Family	Mean age of onset	PS1 mutation	Generations	Affected individuals	CWP	Clinical presentation	References
EB	36	ΔI83/M84	3	5	+	AD + SP	Houlden et al. 2000; Steiner et al. 2001
Japanese	40	Y154N	2	2	n.d.	AD + SP	Hattori et al. (2004)
Michigan	<30	InsFI ex5	4	5	+	AD + SP	Moretti et al. 2004; Rogaeva et al. 2001
M	40	G217D	2	5	+	AD + parkinsonism	Takao et al. 2002
Japanese	31	F237I	1	1	n.d.	AD + SP	Sodeyama et al. 2001
-	38	V261F	2	4	n.d.	AD + SP	Farlow et al. 2000
French	20–54	P264L	3	4	+	Atypical AD + SP	Jacquemont et al. 2002
Tas-1	49	L271V	3	13	+	AD	Kwok et al. 2003
P-2	37	R278T	1	1	–	AD + SP	Kwok et al. 1997
Italian	45	R278K	–	3	n.d.	AD + SP	Assini et al. 2003
Irish	42	E280G	3	9	+	AD + SP	O'Riordan et al. 2002
Canadian	52	E280G	2	5	+	AD + SP	Rogaeva et al. 2003
Japanese	32	P284L	1	1	+	AD + SP	Tabira et al. 2004
EOFAD1	46	Δex9 (5.9 kb)*	3	13	+	AD + SP	Smith et al. 2001
EOFAD2	39	Δex9 (G/A s.a.)*	3	14	+	AD + SP	Kwok et al. 1997
EOFAD3	44	Δex9 (G/T s.a.)*	2	6	+	AD + SP	Brooks et al. 2003
FINN2	50	Δex9 (4.6 kb)*	4	23	+	AD + SP	Crook et al. 1998
FINN3	42.5	Δex9 (4.6 kb)*	2	4	+	AD	Hiltunen et al. 2000

Table 2. *Continued*

Family	Mean age of onset	PS1 mutation	Generations	Affected individuals	CWP	Clinical presentation	References
TK-1	47	Δex9 (G/A s.a.)*	4	12	+	AD + SP	Sato et al. 1998; Mann et al. 2001; Tabira et al. 2002
F74	42	Δex9 (G/T s.a.)*	3	7	+	Dyskinesia + SP	Perez-Tur et al. 1995; Mann et al. 2001
TOR	–	Δex9*	4	14	+	AD + SP	Rogaeva (personal comm.)
Japanese	48	N405S	1	1	+	AD + SP	Yasuda et al. 2000
SYD-1	<30	P436Q	1	1	–	AD + SP	Taddei et al. 1998
D	29	P436Q	2	3	+	AD + SP	Houlden et al. 2000
UK	35	P436Q	2	2	+	AD + SP	Beck et al. 2004
CO	31	n.d.	3	5	+	AD + SP	Houlden et al. 2000;Van Bogaert et al. 1940
Japanese	28	n.d.	1	1	+	AD + SP	Tabira et al. 2002
Japanese	26	n.d.	1	1	+	AD + SP	Tabira et al. 2003

* Sometimes called Δexon10 due to an untranslated exon identified in the 5' end of the gene

tions have been reported. Three Australian AD pedigrees with SP have a deletion of exon 9. EOFAD1 has a 5.9 kb genomic deletion, and EOFAD2 and EOFAD3 have two different splice acceptor mutations (Kwok et al. 1997; Smith et al. 2001; Brooks et al. 2003). A consequence of the loss of PS1 exon 9 is that the truncated transcript also carries an additional amino acid substitution, S290C, arising from the codon spanning the splice site. Steiner et al. (1999) examined the effect of the S290C mutation and showed that this mutation also resulted in increased Aβ production, indicative of a pathogenic mutation.

The EOFAD1 pedigree is a three-generation early-onset pedigree with 13 affected subjects. Members of two generations developed presenile AD, confirmed by autopsy in two cases. Members of one sibship in the third generation, at 50% risk, remained asymptomatic until their mid-40s, an age at which affected family members usually had established cognitive decline. Four siblings subsequently developed SP in their mid-40s or later. One of these individuals, IV:11, has been examined repeatedly and has remained without significant dementia for many years (Fig. 1). He is currently in a nursing home because of severe physical dis-

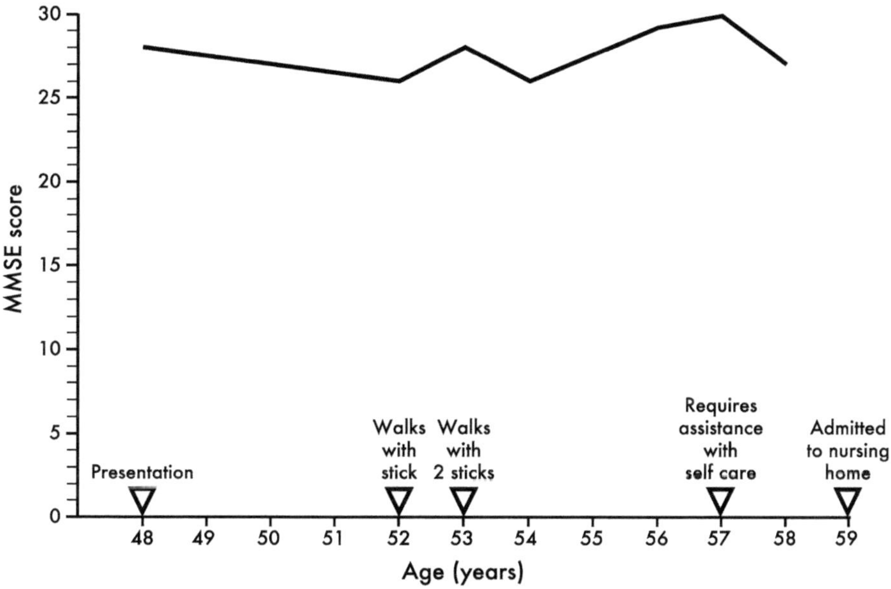

Fig. 1. MMSE score versus age in a EOFAD1 pedigree member with SP. Physical signs of SP documented over time in patient IV:11. Note that MMSE demonstrates that, despite carrying the PS1 exon 9 deletion mutation, the patient has remained dementia free for over 10 years.

ability, but until recently he was able to use a keyboard to conduct telephone conversations, though physically unable to speak owing to motor speech difficulty. A second sibling with SP died prematurely of breast cancer at 53 and was considered not to have dementia. At autopsy, there was pyramidal tract degeneration in the medulla and spinal cord, consistent with SP, but in the cortex there were significant numbers of non-cored plaques, of a type later to be known as "cotton wool" plaques (CWP), with very few tangles and no significant cell loss. An example is seen in Figure 2. Congophilic angiopathy was prominent. A third sibling became cognitively impaired in her 50s at about the same time as she developed SP. Brain MRI showed severe white matter disease, though this was not found in her brother, IV:11, who has remained without significant dementia (Fig. 3). AD with CWP and pyramidal tract degeneration was found at autopsy, together with congophilic angiopathy. Her daughter has developed dementia in her late 40s without SP, indicating that the dementia phenotype can be inherited from a parent with SP as well as vice versa In this family, apart from the striking preservation of cognitive performance in individual IV:11, it is also notable that the mean age at onset of symptoms in individuals with SP at presentation is later than that for individuals with a dementia presentation (Fig. 4).

Brooks et al. (2003) also described the EOFAD3 family with six affected individuals in two generations who carry the exon 9 G/T splice acceptor mutation first described in British family F74 by Perez-Tur et al. (1995). The index case had typical familial AD with onset at 40 years. When examined at 43, she had frontal release signs, positive glabellar tap and moderate rigidity of upper and lower limbs with some cogwheeling but no spasticity. She died at age 45. Her mother and an older sister had been similarly affected, but an older brother, while not considered to have dementia, had developed problems walking at 47 after a fall; a neurologist diagnosed SP. Again, this was a relatively late age of symptom onset. At 51 he was

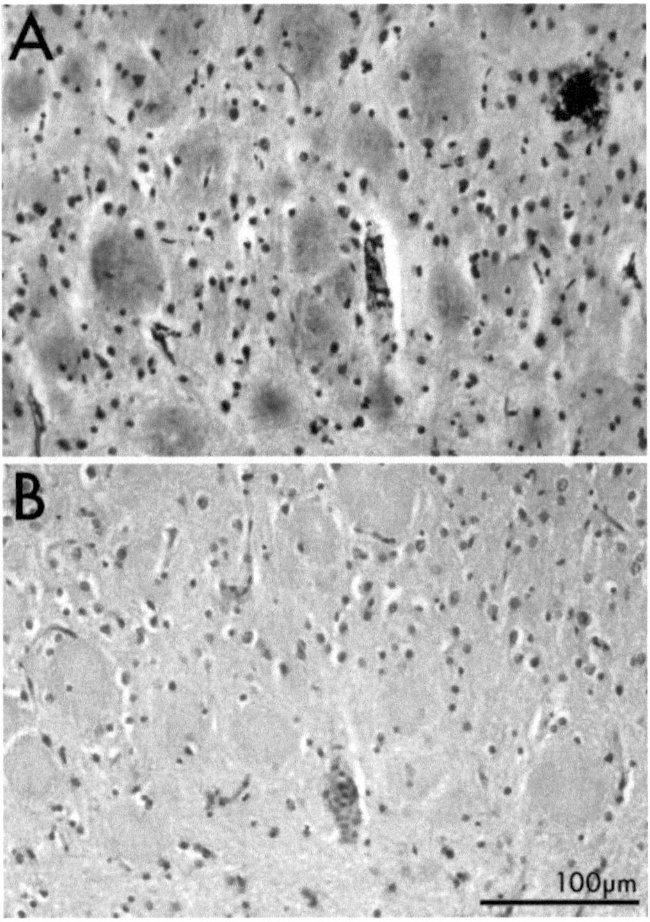

Fig. 2. CWP in the cortex of a patient from the EOFAD1 pedigree. Significant numbers of CWP, with very few tangles and no significant cell loss, were visualised in the cortex with (**A**) silver staining and (**B**) Hematoxylin and eosin (H&E).

Fig. 3. Brain MRI of members of the EOFAD1 pedigree. A–C. Three coronal sections of the brain of a patient with spastic paraparesis and dementia, showing severe white matter disease. Age 60, eight years after onset, MMSE 6/30. D–G. Four transverse sections of the brain of a patient with spastic paraparesis but no dementia. Age 54, six years after onset, MMSE 27/30.

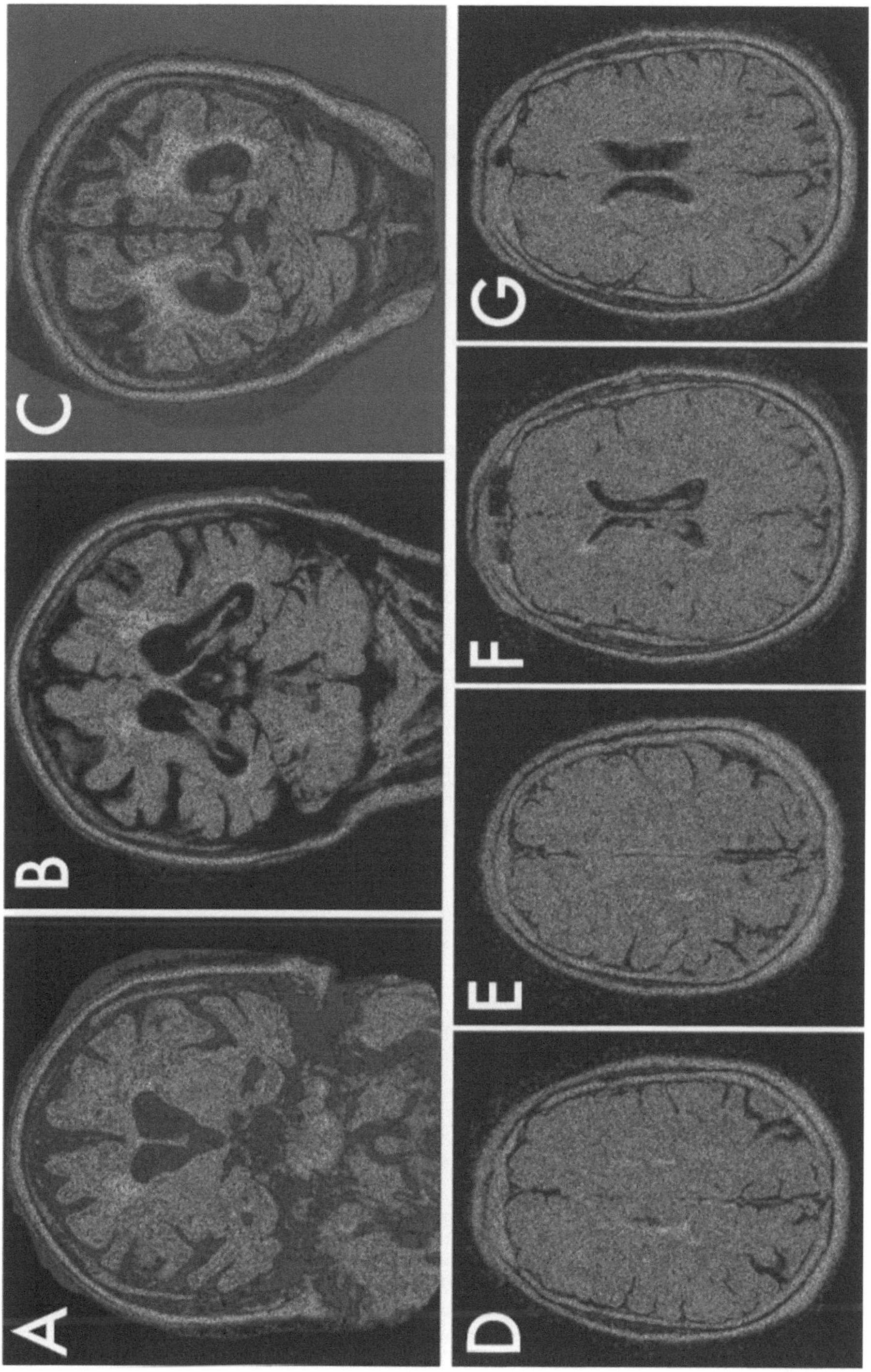
A
B
C
D
E
F
G

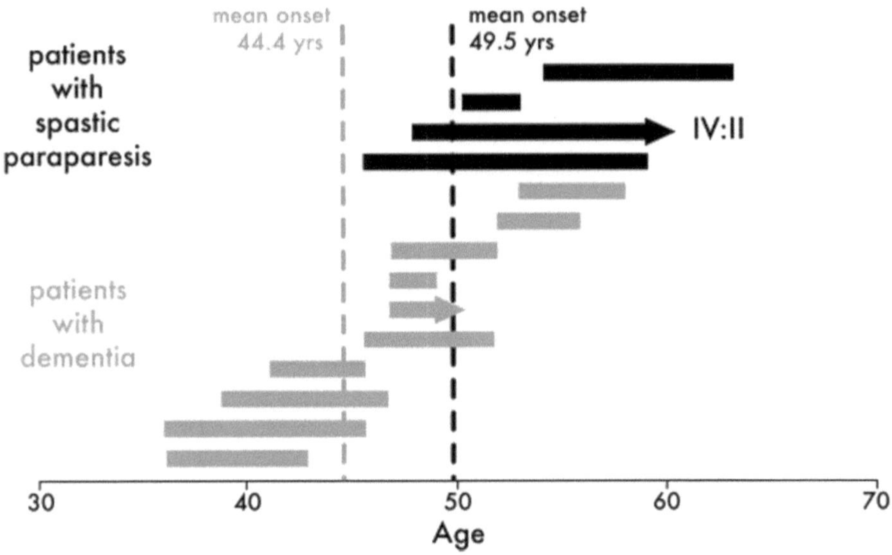

Fig. 4. Onset age and disease duration in members of the EOFAD1 pedigree. Individuals presenting with SP had a later dementia mean onset age (49.5 years) than those presenting with dementia (44.4 years).

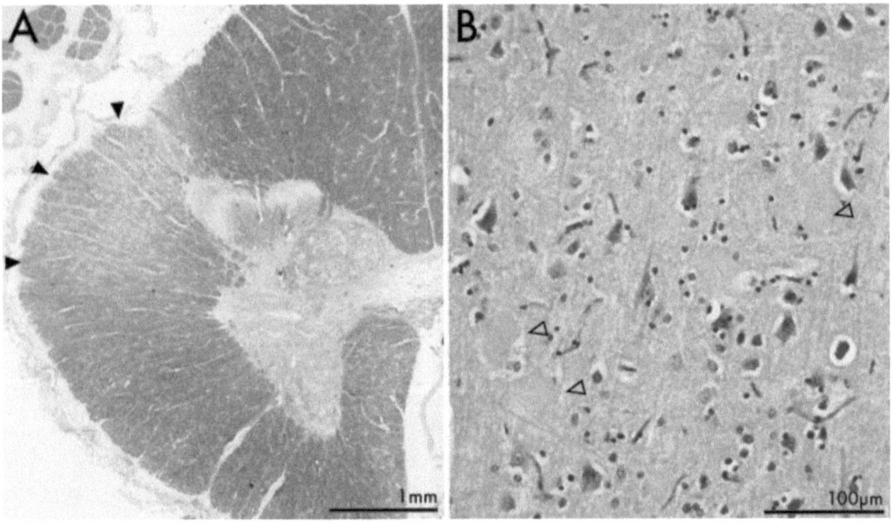

Fig. 5. Neuropathological examination of a patient from the EOFAD3 pedigree. The patient had both dementia and SP. **A.** Myelin staining in the cervical spinal cord showed corticospinal tract degeneration (filled arrowheads) using the Luxol fast blue (LFB) method. **B.** Hematoxylin and eosin (H&E) staining of the motor cortex showed degeneration of the corticospinal cells (open arrowheads).

still living alone; he scored 23/30 on the Mini-Mental State Examination (MMSE; Folstein et al. 1975), losing points for the date and season and for subtraction of serial sevens, though he remembered three objects after an interference task. He had left school at 14 years and claimed he had never learned to write but was able to copy intersecting pentagons. At age 54, he was found to have pseudobulbar palsy; by age 55, he had bladder problems and was in a wheelchair but was still living alone. The following year he was admitted to a nursing home, where he deteriorated over several years with slurred speech and, eventually, complete aphasia, intermittent confusion and agitation, which were worse during episodes of urinary infection and septicaemia. At 58 years, he had several generalised seizures. He recovered but died a few weeks later. Another sibling developed cognitive decline in her mid-40s and, at 47 years of age, she scored 11/30 on the MMSE. Some years later she developed a spastic gait and required a wheelchair. She died at 51, and neuropathological examination of the cervical spinal cord showed bilateral degeneration of the corticospinal tracts (Fig. 5).

The EOFAD2 family has 14 affected subjects in three generations with a deletion of exon 9, this time because of the G/A splice acceptor mutation, first reported in a Japanese family by Sato et al. (1998). Pedigree members had FAD with age at symptom onset ranging from the late 30s to the early 50s. Four affected members who we examined had typical AD without SP; one, however, again with a relatively late onset age of 48 years, developed SP concurrently with cognitive impairment. At 52, he scored 19/30 on the MMSE and, although physically disabled, could drive a motorised wheelchair to the local swimming pool about 1 km from home without getting lost (Brooks et al. 2003).

The Finnish pedigree (FINN2), which is a four-generation pedigree, has 23 affected subjects; 10 of the 14 affected individuals had both AD and SP. This family has a 4.6 kb genomic deletion in the PS1 gene that leads to the loss of exon 9 (Crook et al. 1998; Hiltunen et al. 2000; Prihar et al. 1999; Verkkoniemi et al. 2000, 2001). Co-occurrence of the SP phenotype with AD delayed the onset of dementia symptoms by five years. Neuropathology showed a profusion of CWP (see below) concentrating in the frontal motor cortices and degeneration of the lateral corticospinal tracts. Memory impairment was present in all patients but appeared simultaneously with, or was preceded by, walking difficulty due to SP (in 10 of 14 cases). Impaired fine coordination of hands (9/14) and dysarthria (6/14) in some patients suggested cerebellar involvement. EEG showed intermittent generalised delta-theta activity. Head MRI showed temporal and hippocampal atrophy; PET showed bilateral temporo-parietal hypometabolism. In a branch of this family, FINN3, the clinical and neuropathological phenotype is that of typical AD without indications of SP or CWP (Hiltunen et al. 2000). A Canadian family, the large TOR pedigree with a PS1 exon 9 deletion, has a sibship comprising six affected individuals with variable occurrence of dementia and SP (Dr. E Rogaeva, personal communication).

Mann et al. (2001) described a British family, F74, also described by Perez-Tur et al. (1995), and a Japanese family, TK1, both of which have a splice acceptor mutation causing a deletion of PS1 exon 9. The F74 family has seven affected individuals in three generations, and the F74 proband was said to be dyskinetic and have CWP in the frontal cortex. The Japanese family had 12 affected subjects in four generations, and Tabira et al. (2002) described identical twins from this family: one developed memory disturbances followed by SP and the other reported trembling and neuralgic pain of the left lower leg with an earlier onset of memory impairment. Both brothers had numerous senile plaques in hippocampus and frontal, temporal and parietal cortices. Some plaques were cored but CWP were also demonstrated by hematoxylin-eosin staining (Mann et al. 2001; Perez-Tur et al. 1995; Sato et al. 1998; Tabira et al. 2002).

Despite the predominance of PS1 exon 9 deletion mutations in pedigrees with AD and SP, a number of other PS1 mutations leading to AD/SP have been detected (Table 2). Houlden et al. (2000) reported two families with AD and SP that share the characteristic CWP pathology: 1) the Scottish family EB, with three generations and five affected individuals, had a deletion of two nucleotides and consequently two amino acids in PS1 exon 4 (DelI83/M84). This family was studied in more detail by Steiner et al. (2001), who demonstrated that the deletion caused an increase in Aβ42 generation and reduced activity in Notch signaling, as has been reported for other PS1 mutations (Steiner et al. 2001); and 2) the Belgian family CO, which was first described by Van Bogaert et al. (1940) as family DC, has three generations and five affected subjects without a mutation having been identified. In each of these families, in which SP was a prominent feature, the characteristic pathology of CWP was also observed (Houlden et al. 2000).

Hattori et al. (2004) reported a PS1 Y154N mutation; the proband had gait disturbances that started at 37 years of age and memory difficulty noted by family at 42 years of age. On admission at age 47, she had SP and a MMSE of 16/30. The proband's mother also presented with gait disturbance in her 40s and died at 69 after two years of abnormal behavior and cognitive dysfunction; however, she remained non-demented for at least 10 years after onset of gait disturbance.

Moretti et al. (2004) described a Michigan family with five affected subjects in four generations with a heterozygous 6-nucleotide insertion (TTATAT) at nucleotide position 715 of the PS1 gene, PS1 InsFI located in exon 5. This mutation has also been described by Rogaeva et al. (2001). These insertions cause an unusually aggressive form of AD that maintains the usual regional hierarchy of pathology while extending abnormalities into more widespread brain areas than is typically seen in sporadic AD. The proband had dementia beginning at 28 years of age. By 32, he had dystonic dysarthria, limb dystonia and SP with spontaneous clonus. PET scan showed hypometabolism in the posterior temporoparietal cortex that later encompassed the posterior cingulate, primary motor cortex and frontal association cortices. The involvement of the primary motor cortex was apparent only after SP was clinically noticeable. Both his father and paternal grandfather were affected and died at age 42 and 35, respectively (Moretti et al. 2004; Rogaeva et al. 2001).

Seven different Japanese pedigrees have been reported with both AD and SP, five of which have a mutation in PS1: F237I, N405S, P284L (three pedigrees) and deletion of exon 9 (two pedigrees). Two pedigrees had unknown mutations. All cases had numerous CWP and NFT and SP (Sodeyama et al. 2001; Tabira et al. 2002; Yasuda et al. 2000). The F237I case developed SP at the age of 31, and one year later, mild memory impairment. PET scan demonstrated marked hypometabolism and hypoperfusion in the bilateral temporoparietal areas, including the primary sensory cortex but also in the motor cortex (Sodeyama et al. 2001). He died at age 35.

Farlow et al. (2000) reported a PS1 V261F mutation in which the patient developed leg stiffening at age 38, but no memory impairment. A couple of years later, he developed dementia and was wheelchair bound. A similar syndrome was reported in the subject's father, sister and paternal aunt (Farlow et al. 2000).

Jacquemont et al. (2002) reported a small French family with a PS1 P264L mutation that has four affected individuals in three generations with preceding SP and atypical dementia that is not of the AD type. The patient presented at 54 years of age with gait problems and lower back pain; he developed SP and cognitive impairment (MMSE 18/30 at age 55) but the pattern of cognitive deficits was not typical of AD, being predominantly visuospatial with relative preservation of memory functions. Brain MRI showed mild atrophy. There was a family history of dementia in his sister, who died at age 63, in his mother, who died at age 70 and also had walking difficulty, and in his maternal grandmother (age at death unknown; Jacquemont et al. 2002).

An Irish family with PS1 mutation, E280G, has nine affected individuals in three generations with either AD or both AD and SP. One subject had SP and white matter abnormalities on cranial MRI. A sibling had an internuclear ophthalmoplegia, spastic-ataxic quadriparesis, and CWP with amyloid angiopathy on brain biopsy. Another affected sibling also had MRI white matter abnormalities, though not SP. The authors considered that the MRI findings might reflect an ischemic leukoencephalopathy due to amyloid angiopathy affecting meningocortical vessels (O'Riordan et al. 2002), as has been suggested in individuals with the Dutch APP693 mutation, where pre-symptomatic carriers had white matter changes on MRI (Bornebroek et al. 1996). Our own observations are consistent with this conclusion. We have also seen severe white matter change in a member of EOFAD-1 with both SP and dementia (Fig. 3A-C). The severity of the changes led the radiologist to suggest a diagnosis of hereditary leukoencephalopathy, but long chain fatty acid levels were normal. At autopsy, there was generalised atrophy, with numerous dilated perivascular spaces in the white matter throughout the cerebral hemispheres. On microscopy, there was widening of perivascular spaces, rarefaction of the white matter and thickening of the vessel walls, with amyloid angiopathy in most meningeal and many parenchymal blood vessels. There were also microinfarcts in the occipital cortex, occipital white matter and frontal white matter. These white matter changes were not found consistently, however. Her brother, with SP only, had many fewer severe changes, though the scan was done two years earlier in his illness (Fig. 3D-G). Rogaeva et al. (2003) reported a Canadian family of five affected subjects in two generations. The pedigree has an E280G PS1 mutation, previously reported by O'Riordan et al. (2002), although it was erroneously published as an E280Q mutation (Dr. E. Rogaeva, personal communication). The proband had memory difficulties and weakness in both legs beginning at age 52 (duration of 15 years). Neuropathology showed degeneration of the corticospinal tracts (Rogaeva et al. 2003).

There are two other small Australian families with mutations in the PS1 gene, R278T and P436Q, respectively, that have only one or two affected subjects each (Kwok et al. 1997; Taddei et al. 1998). The R278T mutation was identified in a patient who developed motor problems at 34 and cognitive deficits at 36. He died at age 41 and brain biopsy confirmed AD. His mother did not carry the mutation; she and his siblings were examined and were unaffected. His father died of cancer at 65 without any apparent neurological deficit. The P436Q mutation was identified in an individual who became forgetful in her late 20s. When seen at age 34, she had moderately severe dementia with a MMSE score of 14/30; her gait was abnormal at that stage and she had brisk reflexes with bilateral extensor plantar responses, though tone was normal. As her dementia progressed, she became physically more disabled, with increasing spasticity, pseudobulbar palsy and rigidity; she died at age 40. Her father had died at 41 of what was said to be cerebral atrophy and was known to have been in a wheelchair. The British family D, with two generations and three affected individuals, also has a PS1 P436Q mutation with CWP (Houlden et al. 2000; Taddei et al. 1998).

Beck et al. (2004) reported a family from UK where germ line and somatic mosaicism of the PS1 P436Q mutation was found in a woman who died at age 58 after a 16-year history of progressive parkinsonism, mild SP and dementia. The mutation was not found in DNA from peripheral blood. Neither parent nor siblings were affected but her daughter developed a progressive cerebellar syndrome at age 27, with SP and dementia; she died after 12 years. She was found to carry the P436Q mutation (Beck et al. 2004).

Assini et al. (2003) reported an Italian family with three affected individuals with a PS1 R278K mutation. One member had dementia onset at 48 years of age, one had SP onset at 45 years of age followed by dementia five years later, and one man had SP onset at 41 years of age with only focal memory deficit after 12 years, with a MMSE score of 28/30 (Assini et al. 2003).

There are 139 PS1 mutations (in 277 pedigrees) described to date, of which 17 (in 24 pedigrees) can lead to SP. However, these 139 PS1 mutations are relatively evenly spaced throughout the PS1 gene, with 7% of mutations located in exon 13 to 21% of mutations located in exons 5 and 7. The stark exception is exon 9, which only has 1.4% of PS1 mutations leading to AD. In contrast, AD/SP mutations, which are generally represented between 1 and 2% of all AD mutations in each exon, are seen at 5–6% of AD mutations in exons 8 and 9. Thus, mutations in exons 8 and 9 and especially the exon 9 deletion mutations are markedly overrepresented.

Cotton wool plaques

Some PS1 mutations have been associated with neuropathological lesions that are not typically observed in AD. CWP were first described in the Finnish family, FINN2, with a 4.6 kb genomic deletion of PS1 exon 9, that has 10 of 14 individuals with SP and dementia (see Table 2; Crook et al. 1998). CWP are spherical with defined margins and do not contain neurites as normal dense cored amyloid plaques do. Moreover, PS1 mutations associated with CWP secrete unusually high levels of Aβ1-42, which is typically located in the peripheral part of the plaque. CWP have little plaque-related glia or immunoreactivity for complement components such as C1q, C3d and C9, suggesting a non-inflammatory pathogenesis.

Other pedigrees with different PS1 mutations have also been reported to have CWP (Brooks et al. 2003; Houlden et al. 2000; Kwok et al. 2003; Moehlmann et al. 2002; O'Riordan et al. 2002; Smith et al. 2001; Steiner et al. 2001; Takao et al. 2002; Table 2). It should be noted that all of these cases also represent pedigrees that have been described above as having members with SP. However, there are some sporadic late-onset AD cases without PS1 mutations that have been reported to have CWP (Le et al. 2001), suggesting that CWP arise from excess Aβ production. Although CWP were initially thought to always be co-associated with the SP phenotype, several pedigrees have been described with CWP but without SP.

Kwok et al. (2003) described a Tasmanian family, Tas-1, with a PS1 missense mutation L271V that results in a PS1 transcript isoform that lacks exon 8. The isoform lacking exon 8 resulted in the increased secretion of the Aβ42 peptide, and the variant PS1 molecule did not interact with either the tau or glycogen synthase kinase 3 beta (GSK-3β) proteins, providing an explanation for the lack of neuritic dystrophy in the CWPs found in this family. The Tas-1 family has 13 affected individuals in three generations with clinical features consistent with AD (Kwok et al. 2003).

Takao et al. (2002) reported a Japanese family "M" with five affected subjects from two generations. Two siblings had a clinical phenotype characterised by dementia and parkinsonism with stooped posture, rigidity and bradykinesia. Numerous CWP, senile plaques, NFT and gliosis were found in both patients. Genetic analysis revealed a G217D PS1 mutation (Takao et al. 2002).

Smith et al. (2001) reported that, in the Australian EOFAD1 family, some individuals developed SP without dementia. Neuropathology showed CWP within the cortex in the subjects with SP only or AD/SP compared to subjects with only dementia, who had normal cored plaques. This finding suggests that CWP are associated with SP, either with or without AD (Smith et al. 2001).

Conclusions

Many different PS1 mutations are associated with classic FAD. However, a striking variation in clinical presentation occurs in some AD pedigrees in which some individuals have SP, either preceding, concurrent with, or instead of dementia and with brain pathology characterised by Aβ-positive CWP. The dementia phenotype is unusual in that dementia onset appears to be delayed compared to affected individuals who present with dementia only. The phenotypic variability can be due to specific alleles of the PS1 gene, environmental effects or the involvement of modifier genes. Variations seen in neuropathology and neurological symptoms in PS1 FAD suggest that an unlinked modifier gene, acting in concert with a PS1 mutation, may underlie this phenotypic heterogeneity. Supporting evidence for this proposition includes the fact that AD/SP pedigrees are due to deletions of exon 9 as well as other PS1 mutations, which suggests that allelic heterogeneity may not be a factor in the SP variant of AD. Secondly, the variable presentation of SP and dementia within large sibships carrying identical PS1 mutations suggests that environmental factors may not be the primary determinant.

However, Houlden et al. (2000) have offered another suggestion for the pathogenesis of the SP variant, namely the amount of Aβ42 production, that is, a dosage effect. Transfection experiments revealed that all SP-causing mutations have a larger effect on Aβ42 production than those mutations that lead to the more typical dementia phenotype. This finding was also confirmed by examining the Aβ42 levels in brains of affected cases (Houlden et al. 2000). While this proposal may explain the presence of CWP in AD/SP pedigrees, it does not provide an explanation for the altered sites of neuropathology, such as the degeneration of the motor cortex, which is normally spared in AD, and the degeneration of the corticospinal tracts.

To identify modifier genes in FAD/SP pedigrees, there are some obvious approaches and candidate genes to analyze. First, the common and defined genes in pure SP pedigrees provide candidates for mutation screening. Mutations in three SP genes, spastin, atlastin and paraplegin, account for approximately 50% of all hereditary SP cases. No coding sequence variation was observed in these genes in an AD/SP pedigree (Rogaeva et al. 2003). Similarly, we have screened the coding exons and flanking introns of the ACP33 gene in our FAD/SP pedigrees without detecting any sequence variations (unpublished data). Second, genes implicated in the phosphorylation of the tau protein are strong candidates, as one of the main differences between demented patients with and without SP is the lack of neuritic involvement surrounding the plaques in patients with SP. Genes of interest in this group include glycogen synthase kinase 3 beta (GSK-3β), cyclin-dependent kinase 5 (cdk5), protein kinase A, casein kinase 1a and the cell division cyclin 2 genes. Third, the candidate genes involved in Aβ degradation, such as insulin degrading enzyme (IDE), neprilysin and α2 macroglobulin, may also be involved in disease pathogenesis and the production of CWP. Initial experiments reveal that no polymorphisms in the IDE gene have been found in our FAD/SP pedigrees (unpublished data). Other ways of identifying modifier genes may be to use cDNA microarray analysis of brain tissues from individuals with or without SP. An alternate approach would be to perform linkage analysis on a cohort of pedigrees with AD/SP to determine if there is a genetic (modifier) locus that is co-inherited with the SP phenotype.

Although the exact mechanism by which the AD/SP phenotype arises remains to be determined, we suggest that the most likely explanation is the presence of modifier genes. The importance of identifying modifier genes involved in AD pathogenesis is that it will provide insight into the mechanisms by which organisms modulate biological processes to accommodate the deleterious effects of genetic mutations. In the case of individuals in AD pedigrees with SP, the putative modifier gene is proposed to delay the onset of dementia in carrier individuals, perhaps by preventing neurodegeneration in cortical regions despite deposition of Aβ, which is frequently seen in the form of CWP. New disease therapeutics could be based on mimicking or enhancing the effects of naturally occurring genetic modifiers.

Acknowledgements

Supported by grants from the Australian National Health and Medical Research Council (157209, 113804, 276401) and a fellowship from the Swedish Wenner-Gren Foundation. We thank Heidi Cartwright for assistance with the artwork.

References

Alzheimer A (1907) Uber eine eigenartige Erkrankung der Hirnrinde. Allg Z Psychiat Gerichte Med 64:146–148.
Assini A, Terreni L, Borghi R, Giliberto L, Piccini A, Loqui, D, Fogliarino S, Forloni G, Tabaton M (2003) Pure spastic paraparesis associated with a novel presenilin 1 R278K mutation. Neurology 60:150–151.
Barrett A (1913) A case of Alzheimer's disease with unusual neurological disturbances. J Nervous Mental Disorder 4c:361–374.
Beck J, Poulter M, Campbell TA, Uphill JB, Adamson G, Geddes JF, Revesz T, Davis MB, Wood NW, Collinge J, Tabrizi SJ (2004) Somatic and germline mosaicism in sporadic early-onset Alzheimer's disease. Human Mol Genet 13:1219–1224.
Behan W, Maia M (1974) Strumpell's familial spastic paraplegia: genetics and neuropathology. J Neurol Neurosurg Psychiat 37:8–20.
Bornebroek M, Haan J, van Buchem MA, Lanser JB, de Vries VD, Weerd MA, Zoeteweij M, Roos RA (1996) White matter lesions and cognitive deterioration in presymptomatic carriers of the amyloid precursor protein gene codon 693 mutation. Arch Neurol 53:43–48.
Brooks W, Kwok JB, Kril JJ, Broe GA, Blumbergs PC, Tannenberg AE, Lamont PJ, Hedges P, Schofield PR (2003) Alzheimer's disease with spastic paraparesis and 'cotton wool' plaques: two pedigrees with PS-1 exon 9 deletions. Brain 126:783–791.
Casari G, De Fusco M, Ciarmatori S, Zeviani M, Mora M, Fernandez P, De Michele G, Filla A, Cocozza S, Marconi R, Durr A, Fontaine B, Ballabio A (1998) Spastic paraplegia and OXPHOS impairment caused by mutations in paraplegin, a nuclear-encoded mitochondrial metalloprotease. Cell 93:973–983.
Chartier-Harlin MC, Parfitt M, Legrain S, Perez-Tur J, Brousseau T, Evans A, Berr C, Vidal O, Roques P, Gourlet V (1994) Apolipoprotein E, epsilon 4 allele as a major risk factor for sporadic early-and late-onset forms of Alzheimer's disease: analysis of the 19q13.2 chromosomal region. Human Mol Genet 3:569–574.
Crook R, Verkkoniemi A, Perez-Tur J, Mehta N, Baker M, Houlden H, Farrer M, Hutton M, Lincoln S, Hardy J, Gwinn K, Somer M, Paetau A, Kalimo H, Ylikoski R, Poyhonen M, Kucera S, Haltia M (1998) A variant of Alzheimer's disease with spastic paraparesis and unusual plaques due to deletion of exon 9 of presenilin 1. Nature Med 4:452–455.
De Strooper B, Annaert W, Cupers P, Saftig P, Craessaerts K, Mumm JS, Schroeter EH, Schrijvers V, Wolfe MS, Ray WJ, Goate A, Kopan R (1999) A presenilin-1-dependent gamma-secretase-like protease mediates release of Notch intracellular domain. Nature 398:518–522.
Engert JC, Dore C, Mercier J, Ge B, Betard C, Rioux JD, Owen C, Berube P, Devon K, Birren B, Melancon SB, Morgan K, Hudson TJ, Richter A (1999) Autosomal recessive spastic ataxia of Charlevoix-Saguenay (ARSACS): high-resolution physical and transcript map of the candidate region in chromosome region 13q11. Genomics 62:156–164.
Farlow M, Murrell JR, Hulette CM, Ghetti B (2000) Hereditary lateral sclerosis and Alzheimer disease associated with mutation at codon 261 of the presenilin 1 (PS1) gene. Neurobiol Aging 21:S62.
Fink J (2003) The hereditary spastic paraplegias: nine genes and counting. Arch Neurol 60:1045–1049.

Folstein M, Folstein SE, McHugh PR (1975) "Mini-mental state." A practical method for grading the cognitive state of patients for the clinician. J Psychiatr Res 12:189–198.

Grundke-Iqbal I, Iqbal K, Quinlan M, Tung YC, Zaidi MS and Wisniewski HM (1986) Microtubule-associated protein tau. A component of Alzheimer paired helical filaments. J Biol Chem 261:6084–6089.

Harding A (1983) Classification of the hereditary ataxias and paraplegias. Lancet 1:1151-1155.

Harding A (1993) Hereditary spastic paraplegias. Sem Neurol 13:333–336.

Hardy J (1997) Amyloid, the presenilins and Alzheimer's disease. Trends Neurosci 20: 154-159.

Hattori S, Sakuma K, Wakutani Y, Wada K, Shimoda M, Urakami K, Kowa H, Nakashima K (2004) A novel presenilin 1 mutation (Y154N) in a patient with early onset Alzheimer's disease with spastic paraparesis. Neurosci lett 386:319–322

Hedera P, DiMauro S, Bonilla E, Wal, J, Eldevik OP, Fink JK (1999) Phenotypic analysis of autosomal dominant hereditary spastic paraplegia linked to chromosome 8q. Am J Human Genet 64:563–569.

Herreman A (2000) Total inactivation of γ-secretase activity in presenilin-deficient embryonic stem cells. Nature Cell Biol 2:461–462.

Hiltunen M, Helisalmi S, Mannermaa A, Alafuzoff I, Koivisto AM, Lehtovirta M, Pirskanen M, Sulkava R, Verkkoniemi A, Soininen H (2000) Identification of a novel 4.6-kb genomic deletion in presenilin-1 gene which results in exclusion of exon 9 in a Finnish early onset Alzheimer's disease family: an Alu core sequence-stimulated recombination. Eur J Human Genet 8:259–266.

Houlden H, Baker M., McGowan, E, Lewis P, Hutton M, Crook R, Wood NW, Kumar-Singh S, Geddes J, Swash M, Scaravilli F, Holton JL, Lashley T, Tomita T, Hashimoto T, Verkkoniemi A, Kalimo H, Somer M, Paetau A, Martin JJ, Van Broeckhoven C (2000) Variant Alzheimer's disease with spastic paraparesis and cotton wool plaques is caused by PS-1 mutations that lead to exceptionally high amyloid-β concentrations. Ann Neurol 48:806–808.

Hughes C, Byrne PC, Webb S, McMonagle P, Patterson V, Hutchinson M, Parfrey NA (2001) SPG15, a new locus for autosomal recessive complicated HSP on chromosome 14q. Neurology 56:1230–1233.

Jacquemont M-L, Campion D, Hahn V, Tallaksen C, Frebourg T, Brice A, Durr A (2002) Spastic paraparesis and atypical dementia cuase by PSEN1 mutation (P264L), responsible for Alzheimer's disease. J Med Genet 39:e2.

Kwok J, Taddei K, Hallupp M, Fisher C, Brooks W, Broe G, Hardy J, Fulham M., Nicholson G, Stell R, St George-Hyslop P, Fraser P. Kakulas B, Clarnette R, Relkin N, Gandy S, Schofield P, Martins R (1997) Two novel (M233T and R278T) presenilin-1 mutations in early-onset Alzheimer's disease pedigrees and preliminary evidence for association of presenilin-1 mutations with a novel phenotype. NeuroReport 8:1537–1542.

Kwok J, Halliday GM, Brooks WS, Dolios G, Laudon H, Murayama O, Hallupp M, Badenhop RF, Vickers J, Wang R, Naslund J, Takashima A, Gandy SE, Schofield, PR (2003) Presenilin-1 mutation L271V results in altered exon 8 splicing and Alzheimer's disease with non-cored plaques and no neuritic dystrophy. J Biol Chem 278:6748–6754.

Le T, Crook R, Hardy J, Dickson DW (2001) Cotton wool plaques in non-familial late-onset Alzheimer disease. J Neuropathol Exp Neurol 60:1051–1061.

Levy-Lahad E, Wasco W, Poorkaj P, Romano DM, Oshima J, Pettingell WH, Yu C, Jondro PD, Schmidt SD, Wang KB, Crowly AC, Fu Y-H, Guenette SY, Galad D, Nemens E, Wijsman EM, Bird TD, Schellenberg GD, Tanzi RE (1995a) Candidate gene for the chromosome 1 familial Alzheimer's disease locus. Science 269: 973–977.

Levy-Lahad E, Wijsman EM, Nemens E, Anderson L, Goddard KAB, Weber JL, Bird TD, Schellenberg, GD (1995b) A familial Alzheimer's disease locus on chromosome 1. Science. 269:970–973.

Mann D, Takeuchi A, Sato S, Cairns NJ, Lantos PL, Rossor MN, Haltia M, Kalimo H, Iwatsubo T (2001) Cases of Alzheimer's disease due to deletion of exon 9 of the presenilin-1 gene show

an unusual but characteristic β-amyloid pathology known as 'cotton wool' plaques. Neuropathol Appl Biol 27:189–196.
Moehlmann T, Winkler E, Xia X, Edbauer D, Murrell J, Capell A, Kaether C, Zheng H, Ghetti B, Haass C, Steiner H (2002) Presenilin-1 mutations of leucine 166 equally affect the generation of the Notch and APP intracellular domains independent of their effect on Aβ 42 production. Proc Natl Acad Sci USA 99:8025–8030.
Moretti P, Lieberman AP, Wilde EA, Giordani BI, Kluin KJ, Koeppe RA, Minoshima S. Kuhl DE, Seltzer WK, Foster NL (2004) Novel insertional presenilin 1 mutation causing Alzheimer's disease with spastic paraparesis. Neurology 62:1865–1868.
O'Riordan S, McMonagle P, Janssen JC, Fox NC, Farrell M, Collinge J, Rossor MN, Hutchinson M (2002) Presenilin-1 mutation (E280G), spastic paraparesis, and cranial MRI white-matter abnormalities. Neurology 59:1108–1110.
Patel H, Cross H, Proukakis C, Hershberger R, Bork P, Ciccarelli FD, Patton MA, McKusick VA, Crosby AH (2002) SPG20 is mutated in Troyer syndrome, an hereditary spastic paraplegia. Nature Genet 4:347–348.
Perez-Tur J, Froelich S, Prihar G, Crook R, Baker M, Duff K, Wragg M, Busfield F, Lendon C, Clark RF, Roques P, Fuldner RA, Johnston J, Cowburn R, Forsell C, Axelman K, Lilius L, Houlden H, Karran E, Roberts GW, Rossor M, Adams (1995) A mutation in Alzheimer's disease destroying a splice acceptor site in the presenilin-1 gene. NeuroReport 7:297–301.
Price DL and Sisodia SS (1998) Mutant genes in familial Alzheimer's disease and transgenic models. Annu Rev Neurosci 21:479–505.
Prihar G, Verkkoniemi A, Perez-Tur J, Crook R, Lincoln S, Houlden H, Somer M, Paetau A, Kalimo H, Grover A, Myllykangas L, Hutton M, Hardy J, Haltia M (1999) Alzheimer disease PS-1 exon 9 deletion defined. Nature Med 5:1090.
Rogaev EI, Sherrington R, Rogaeva EA, Levesque G, Ikeda M, Liang Y, Chi H, Lin C, Holman K, Tsud, T, Mar L, Sorbi S, Nacmias B, Piacentini S, Amaducci L, Chumakov I., Cohen D, Lannfelt L, Fraser PE, Rommens JM and St George-Hyslop PH (1995) Familial Alzheimer's disease in kindreds with missense mutations in a gene on chromosome 1 related to Alzheimer's disease type 3 gene. Nature 376:775–778.
Rogaeva E, Fafel KC, Song YQ, Medeiros H, Sato C, Liang Y, Richard E, Rogaev EI Frommelt P, Sadovnick AD, Meschino W, Rockwood K, Bos, MA, Mayeux R, St George-Hyslop P (2001) Screening for PS1 mutations in a referral-based series of AD cases: 21 novel mutations. Neurology 57:621–625.
Rogaeva E, Bergeron C, Sato C, Moliak I, Kawarai T, Toulina A, Song YQ, Kolesnikova T, Orlacchio A, Bernardi G, St George-Hyslop PH (2003) PS1 Alzheimer's disease family with spastic paraplegia: the search for a gene modifier. Neurology 61:1005–1007.
Sato S, Kamino K, Miki T, Doi A, Ii K, St George-Hyslop PH, Ogihara T, Sakaki Y (1998) Splicing mutation of presenilin-1 gene for early-onset familial Alzheimer's disease. Human Mutat Suppl 1:S91–S94.
Schwarz G, Liu CN. (1956) Hereditary (familial) spastic paraplegia. Further clinical and pathologic observations. AMA Arch Neurol Psychiatr 75:144–162.
Selkoe DJ (1994) Cell biology of the amyloid β-protein precursor and the mechanism of Alzheimer's disease. Annu Rev Cell Biol 10:373–403.
Simpson M, Cross H, Proukakis C, Pryde A, Hershberger R, Chatonnet A, Patton MA, Crosby AH (2003) Maspardin is mutated in mast syndrome, a complicated form of hereditary spastic paraplegia associated with dementia. Am J Human Genet 73:1146–1156.
Smith M, Kwok JBJ, McLean CA, Kril JJ, Broe GA, Nicholson GA, Cappai R, Hallupp M, Cotton RGH, Masters CL, Schofield PR, Brooks WS (2001) Variable phenotype of Alzheimer's disease with spastic paraparesis. Ann Neurol 49:125–129.
Sodeyama N, Iwata T, Ishikawa K, Mizusawa H, Yamada M, Itoh Y, Otomo E, Matsushita M, Komatsuzaki Y (2001) Very early onset Alzheimer's disease with spastic paraparesis associated with a novel presenilin 1 mutation (Phe237Ile). J Neurol Neurosurg Psychiat 71:556–557.

Steiner H, Romig H, Grim MG, Philipp U, Pesold B, Citron M, Baumeister R, Haass C (1999) The biological and pathological function of the presenilin-1 Deltaexon 9 mutation is independent of its defect to undergo proteolytic processing. J Biol Chem 274:7615–7618.

Steiner H, Revesz T, Neumann M, Romig H, Grim MG, Pesold B, Kretzschmar HA, Hardy J, Holton JL, Baumeister R, Houlden H, Haass C (2001) A pathogenic presenilin-1 deletion causes aberrant Aβ 42 production in the absence of congophilic amyloid plaques. J Biol Chem 276:7233–7299.

Tabira T, Chui DH, Nakayama H, Kuroda S, Shibuya M (2002) Alzheimer's disease with spastic paresis and cotton wool type plaques. J Neurosci Res 70:367–372.

Taddei K, Kwok JBJ, Kril JJ, Halliday GM, Creasey H, Hallupp M, Fisher C, Brooks WS, Chung C, Andrews C, Masters CL, Schofield PR, Martins RN (1998) Two novel presenilin-1 mutations (Ser169Leu and Pro436Gln) associated with very early onset Alzheimer's disease. NeuroReport 9:3335–3339.

Takao M, Ghetti B, Hayakawa I, Ikeda E, Fukuuchi Y, Miravalle L, Piccardo P, Murrell JR, Glazier BS, Koto A (2002) A novel mutation (G217D) in the Presenilin 1 gene (PSEN1) in a Japanese family: presenile dementia and parkinsonism are associated with cotton wool plaques in the cortex and striatum. Acta Neuropathol 104:155–170

van Bogaert L, Maere M, de Smedt E (1940) Sur les formes familiales précoces de la maladie d'Alzheimer. Monatsschrift Psychiatr Neurol 102:249–301.

Verkkoniemi A, Somer M, Rinne JO, Myllykangas L, Crook R, Hardy J, Viitanen M, Kalimo H, Haltia M (2000) Variant Alzheimer's disease with spastic paraparesis: clinical characterization. Neurology 54:1103–1109.

Verkkoniemi A, Kalimo H, Paetau A, Somer M, Iwatsubo T, Hardy J, Haltia M (2001) Variant Alzheimer disease with spastic paraparesis: neuropathological phenotype. J Neuropathol Exp Neurol 60:483–492.

Yasuda M, Maeda S, Kawamata T, Tamaoka A, Yamamoto Y, Kuroda S, Maeda K, Tanaka C (2000) Novel presenilin-1 mutation with widespread cortical amyloid deposition but limited cerebral amyloid angiopathy. J Neurol Neurosurg Psychiat 68: 220–223.

Presenilin Mutations: Variations in the Behavioral Phenotype with an Emphasis on the Frontotemporal Dementia Phenotype

Neill R Graff-Radford[1]

Summary

The vast majority of families with presenilin (PSEN) mutations have the clinical phenotype of Alzheimer's disease. However, there are reports of patients who carry PSEN mutations and have Alzheimer's disease with a variety of other clinical phenotypes including spastic paraplegia, seizures, myoclonus, parkinsonism, epilepsy and amyloid angiopathy. Remarkably, three of the studied families have frontotemporal dementia (FTD). The mutations associated with FTD are L113P (Raux et al. 2000), Ins R362 (Tang-Wai et al. 2002) and G183V (Dermaut et al. 2004). Raux and colleagues (2000) reported six members from four generations of the SAL family who had early onset FTD. Tang-Wai and colleagues (2002) reported three patients from three generations who had FTD in their 50s and 60s. Dermaut and colleagues (2004) reported a large family from two generations. Three were definitely affected and a total of 12 members were evaluated. Again the disease was of early onset. In all three families, the clinical phenotype was convincingly FTD in nature. In the first two families (L113P and Ins R362), no autopsy was available, but in the third family (G183V), one case had an autopsy and the pathology showed Pick's disease with Pick bodies and no Alzheimer pathology. Usually PSEN1 mutations enhance the γ-secretase effect on the amyloid precursor protein (APP), increasing Aβ42 protein, but a study by Amtul and colleagues (2002) found that the InsR 362 (but not L113P, which they also tested) caused a "dominant negative" effect on the metabolism of APP (and NOTCH), decreasing Aβ42 production. The G183V mutation does not have the same effect. In this family, there are two siblings without the mutation (II-3 age 67and II-4 age 66) who had abnormal SPECT scans and mild dysexecutive function, and II-3 had anomia and mild MRI atrophy. All of these findings raise the possibility that FTD might not be linked to the G183V mutation. In conclusion, in the families described so far, there is suggestive but not conclusive evidence that PSEN1 mutations can cause FTD.

[1] Professor of Neurology, Mayo College of Medicine, Department of Neurology, Mayo Clinic Jacksonville, 4500 San Pablo Road, Jacksonville, FL 32224, USA

Cummings et al.
Genotype – Protéotype – Phenotype
Relationships in Neurodegenerative Diseases

Introduction

Three mutations in the Presenilin 1 (PSEN1) gene have been associated with the clinical presentation of frontotemporal dementia (FTD; Raux et al. 2000; Tang-Wai et al. 2002; Dermaut et al. 2004; Fig. 1). In determining if the PSEN1 mutations cause FTD, the evidence we shall evaluate includes the clinical findings, linkage, a feasible pathogenic mechanism and the pathological findings. We should, however, accept that all of these factors so far provide circumstantial evidence and not conclusive proof that these mutations cause FTD.

As background, Sherrington and colleagues (1995) and then Levy-Lehad and colleagues (1995) reported that mutations of PSEN1 and PSEN2, respectively, cause early onset Alzheimer's disease (AD). Scheuner and colleagues (1966) showed that mutations in both of these genes and the amyloid precursor protein (APP) gene increased plasma Aβ42 levels, supporting the hypothesis that early onset AD (EOAD) is caused by a common mechanism: increased Aβ production. Later it was found that PSEN1 and PSEN2 are membrane proteins that are thought to be part of the γ secretase complex, an aspartyl protease system involved in APP processing (Kimberley et al. 2003). γ-Secretase acts not only on APP but also on other type 1 transmembrane proteins, such as Notch 1–4 and the Erb-4 receptor tyrosine kinase (Kimberley et al. 2003). In some of γ-secretase actions, intracellular domains are created that are involved in signaling events (Fortini 2002). In trying to understand how PSEN1 mutations may cause FTD, we should note that PSEN1 gene knockout mice results in embryonic death (Shen et al. 1997) and that conditional ablation of PSEN1 and PSEN2 expression in the brain of adult mice has been shown to result in neurodegeneration and cognitive impairment (Saura et al. 2004). Further, some artificial mutations introduced into PSEN1 (D257E, D385E and delTM1–2; Amtul et al. 2002) have a dominant-negative effect, resulting in loss of γ-secretase activity. In contrast to the increased production of Aβ42 associated with the AD-related PSEN1 mutations, a question arises: could some or all of the FTD-associated mutations be related to a dominant-negative effect on γ-secretase activity, resulting in degeneration?

Review of the reports

L113P Family

The first FTD family with a PSEN1 mutation was reported by Raux and colleagues (2000). The PSEN1 mutation is L113P.

Clinical findings

Family SAL of France included six affected members with EOAD in four generations. Three had a good history for analysis. The first case was a woman with onset at 49 who had personality change, defective judgment, impulsive spending and stereotypic behavior. After nine years she became mute. Computerized tomog-

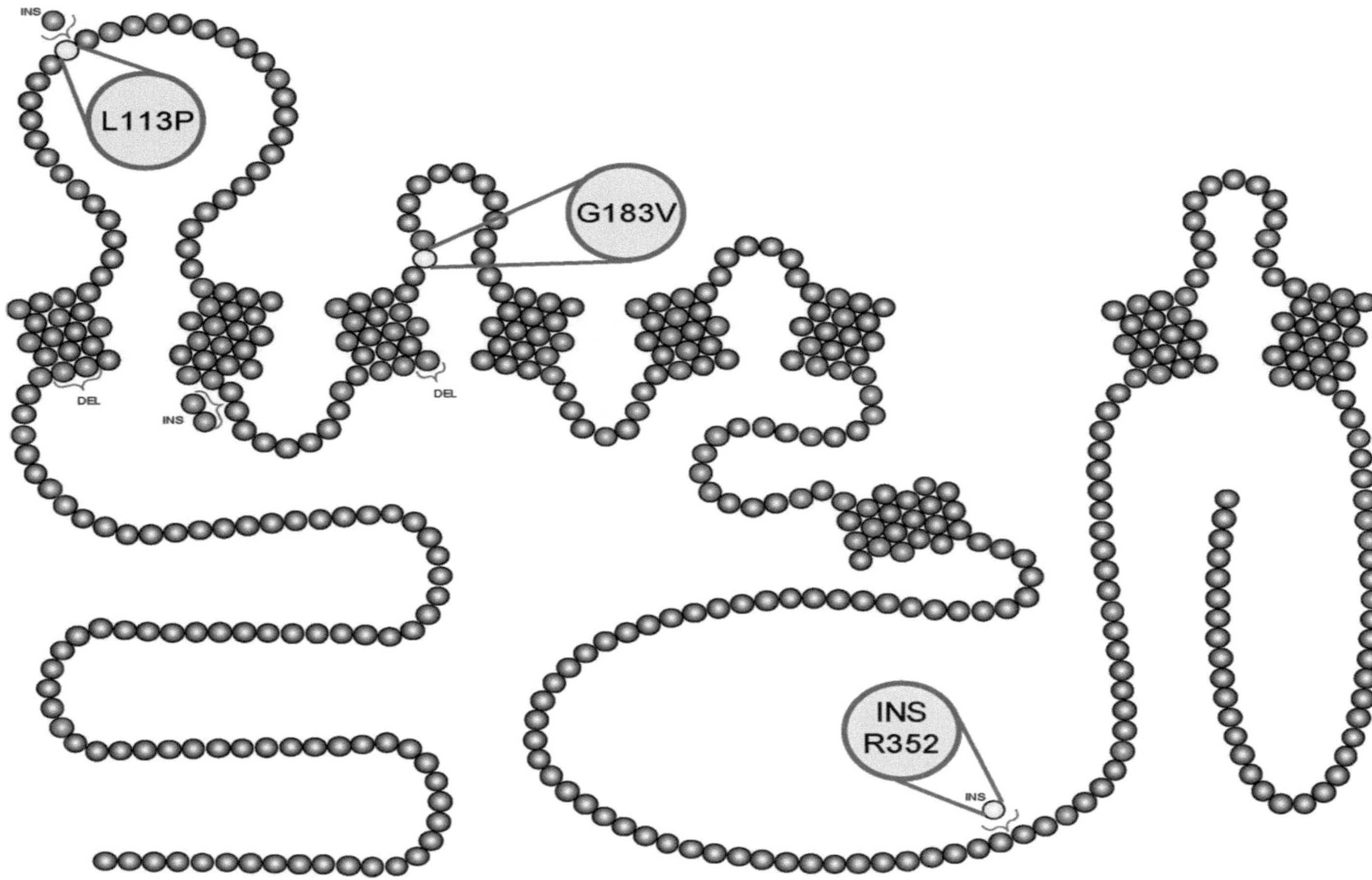

Fig. 1. Mutations in the presenilin 1 gene that have been associated with the clinical presentation of frontotemporal dementia.

raphy (CT) showed frontotemporal atrophy. The second case was a 39-year-old woman who presented with hyperorality and gained 10 Kg. She lacked awareness of her symptoms, was not amnestic, had preserved spatial orientation but had a reduction of speech and eventually was mute. She died 11 years after onset but no autopsy was performed. The third case, a 40-year-old woman had logorrhea, apathy, and neglect of personal hygiene but preserved spatial orientation. On the Mini Mental Status examination (MMSE; Folstein et al. 1975), she scored 19/30, on the Mattis Dementia Rating Scale (DRS; Mattis 1983) 102/144, and on the Boston Naming Test54/60 (Goodglass et al. 1983). She was impaired on the Verbal Fluency (Benton and Hamsher 1983), Trails B (Reitan and Wolfson 1985), Wisconsin Card Sorting (Heaton 1981) and Go-No-Go Test. Her CT scan showed frontal atrophy and her Single Photon Emission Computerized Tomographic scan (SPECT) showed frontal hypoperfusion.

DNA analysis

Two affected siblings had the L113P mutation and one unaffected sibling did not.

This mutation occurs in the hydrophilic loop of PSEN1, in which several mutations are found.

The mutation was not found in 50 unrelated persons.

Pathology

According to personal communication with Dr. Campion in 2004, no autopsy has been performed on any family member.

Proposed pathogenesis

The authors speculate that the pathology will be AD with prominent frontotemporal plaque and tangle pathology.

InsR362 family

Clinical findings

The next family with an FTD phenotype and a PSEN1 mutation was reported by Tang-Wai and colleagues (2002). They described three affected individuals in three generations. The paternal grandfather developed forgetfulness and violent behavior starting in his 60s and died in a nursing home in his 70s. The father had forgetfulness in his 60s and, a short while later, behavioral dyscontrol. The proband, a retired teacher, had hypersomnolence and forgetfulness at age 56. She became lost while driving and ate large amounts of food. On the Short Test of

Mental Status)Kokmen et al. 1991), she scored 32/38 and on formal neuropsychological testing she was impaired in attention, verbal fluency, learning, retention, visuopatial tasks and problem solving. While being tested, the examiner noted that she drew smiley faces and clover-leaves on her tests. She had a voracious appetite, developed delusions and had behavioral dyscontrol. Magnetic Resonance Imaging (MRI) showed prominent frontotemporal atrophy.

Genetic analysis

Her DNA showed a three base pair insertion after nucleotide 1055 at codon 352 in exon 10, resulting in a frame shift with an arginine insertion.

Pathology

No autopsy has been performed.

Proposed pathogenesis
Amtul and colleagues (2002) reported an elegant study showing that this mutation causes a dominant-negative effect, resulting in inhibition of γ-secretase cleavage of APP and Notch. While there are animal data showing that γ-secretase inhibition leads to brain degeneration (Saura et al. 2004), exactly how this would lead to frontotemporal degeneration is unclear.

G183V family

Clinical findings

Dermaut and colleagues (2004) described the proband designated case II:2. The age of onset was 52 and the patient was evaluated at 54. At this time, he had personality change, loss of initiative, apathy, emotional blunting, disinhibition, euphoria, inappropriate laughing, and dietary hyperactivity but had preserved memory and orientation. On examination, he had palmomental, snout and glabella reflexes but no grasp reflexes. Neuropsychological testing revealed a MMSE (Folstein et al. 1975) of 26/30 and he scored in the normal range on the Wechsler Memory Scale (WMS; Wechsler 1987). His testing showed impairment on the Verbal Fluency Test (Benton and Hamsher 1983), proverb interpretation and problem solving. The examiner also observed perseverations, utilization behavior, echolalia and stereotypic behaviors. The patient was carefully followed and, by age 55, hardly spoke and had prominent grasp reflexes. At age 59, he scored 2/30 on the MMSE and 0/60 on the Boston Naming Test (Goodglass et al. 1983). He was bedridden at 61 and died at 62.

Other family members of G183V family

Eleven additional family members, including six siblings, gave DNA and were evaluated.

The proband's father had a 23-year, slowly progressive degenerative dementia and died at 77. His history is compatible with FTD. The proband's eldest brother committed suicide at age 55. At this time, he had a nine-year history in keeping with FTD.

Siblings of the proband without the mutation

Other siblings were evaluated and three did not have the mutation.

- Sibling II-3 at age 67 had an abnormal SPECT, mild atrophy on MRI, severe word-finding difficulty, moderate executive dysfunction and mild memory problems.
- Sibling II-4 at 66 had an abnormal SPECT, normal MRI and mild executive dysfunction.
- Sibling II-5 was normal on evaluation.

Siblings of the proband with the mutation

- Sibling II-6 at 60 had marked hypoperfusion on the SPECT, atrophy on the MRI and a moderate global amnestic syndrome.
- Sibling II-7/8 at 58 had discrete hypoperfusion on the SPECT and mild dysexecutive syndrome and one episode of depression.
- Sibling II-9 at age 56 had a normal SPECT, MRI and mild word-finding difficulty, mild dysexecutive syndrome and mild memory problems.

Pathology

Proband II-2 came to autopsy and macroscopic examination showed that there was severe frontotemporal atrophy with relative sparing of the superior temporal gyrus. The ventricles were dilated. Histological examination showed severe neuronal loss in the frontal, temporal (except superior temporal), hippocampal and cingulate gyri. Fibrillary gliosis was also noted. With Bodian stain, argyrophilic neuronal bodies (Pick's bodies) were present in lamina II and III in the deeper layers of the superior frontal, superior temporal and middle temporal gyri. Pick's bodies were also found in the dentate fascia, CA4 through CA1 zones of the hippocampus, locus ceruleus, periaqueductal gray matter and the griseum pontis. Some of the cells were ballooned, reminiscent of Pick's cells. The inclusions stained with AT8 directed against phosphorilated tau. In Pick's cells and in large neurons, AT8 stained the cytoplasm. Antiubiquitin antibodies stained the inclusions less intensely. They found no amyloid plaque pathology. Analysis of the sarcosyl insoluble tau pellets showed strong doublets at 60 and 64 kDa in the

Table 1. Summary of findings in the three families

Mutation	FTD phenotype	Clinical/ Genetic Association	Pathology	Proposed pathogenesis
L113P	Yes	2 cases	No	AD pathology in frontal lobes
InsR362	Yes	1 case	No	Dominant negative
G183V	Yes	2 cases, 1 possibly not	Yes	Unknown

proband and FTD control (in keeping with three-repeat tau isoforms) whereas, in a PSEN1 mutation patient with AD and an AD patient, there was an additional 69kDa band (in keeping with four-repeat tau isoforms).

Proposed pathogenesis

The pathogenesis is unknown. A summary of the three families is shown in Table 1.

Discussion

The clinical findings of FTD in these three families with PSEN1 mutations (L113P, insR352 and G183V; Raux et al. 2000; Tang-Wai et al. 2002; Dermaut et al. 2004) suggest that PSEN1 mutations may lead not only to AD but also, in some instances, to FTD. The interesting findings of the dominant-negative effect of insR352 mutation inhibiting γ-secretase activity (Amtul et al. 2002) and that conditional ablation of PSEN1 and PSEN2 expression in the brain of adult mice has been shown to result in neurodegeneration and cognitive impairment (Shen et al. 1997; Saura et al. 2004) are important avenues to explore. However, this does not seem to be the mechanism in the other two mutations. As proposed by Raux et al. (2000) authors of the L113P mutation report, AD pathology predominantly in the frontal lobes may lead to an FTD picture. Neither of these two proposed mechanisms can explain the findings in the G183V mutation, in which an autopsy of one case clearly shows Pick's disease and no dominant-negative effect (the same as the insR362 mutation). An unresolved issue in this family is how the clinical picture will evolve in the six siblings. Will those without the mutation develop FTD and Pick's disease or will only those with the mutation develop the clinical picture? In other words, will there be true linkage between the mutation and the clinical phenotype?

In conclusion, we should be careful in interpreting the findings in these three families. Linkage has not been definitively established in any of the families, so it is possible that these mutations are rare benign variants (although all were absent in control populations). Further pathology in these families and discovery of new families may help clarify this interesting and evolving field.

Acknowledgments

Dr Richard Crook PhD made the figure. This paper was partially supported by a grant from Clarice and Robert Smith.

References

Amtul Z, Lewis PA, Piper S, Crook R, Baker M, Findlay K, Singleton A, Hogg M, Younkin L, Younkin SG, Hardy J, Hutton M, Boeve BF, tang-Wai D, Golde TE (2002) A presenilin 1 mutation associated with familial frontotemporal dementia inhibits gamma-secretase cleavage of APP and notch. Neurobiol Dis 9:269–273.

Benton AL, Hamsher K (1983) Multilingual aphasia examination. Iowa City: AJA Associates

Dermaut B, Kumar-Singh S, Engelborghs S, Theuns J, Rademakers R, Saerens J, Pickut BA, Peeters K, van den Broeck M, vennekens K, Claes S, Cruts M, Cras P, Martin JJ, Van Broeckhoven C, De Deyn PP (2004) A Novel presenilin 1 mutation associated with Pick's disease but not β-amyloid plaques. Ann Neurol 55:617–626.

Folstein MF, Folstein SE, McHugh PR (1975) "Mini-Mental State": a practical method for grading the cognitive state of patients for the clinician. J Psychiat Res 12:189–198.

Fortini M (2002). Gamma-secretase-mediated proteolysis in cell-surface-receptor signalling. Nature Rev Mol Cell Biol 3:673–684.

Goodglass H, Kaplan E, Weintraub (1983). The Boston naming test. Ed 2. Philadelphia: Lea & Febiger.

Heaton R (1981) Wisconsin card sorting test. Odessa, Fl: Psychological Assessment Resources

Kimberly WT, LaVoie MJ, Ostaszewski BL, Ye W, Wolfe MS, Selkoe DJ (2003) Gamma-secretase is a membrane protein complex comprised of presenilin, nicastrin, Aph-1, and Pen-2. Proc Natl Acad Sci USA 100:6382–6387.

Kokmen E, Smith G, Petersen R, Tangalos E, Ivnik R (1991) The short test of mental status: Correlations with standardized psychometric testing. Arch Neurol 48:725–728.

Levy-Lahad E, Wasco W, Poorkaj P, Romano DM, Oshima J, Pettingell WH, Yu CE, Jondro PD, Schmidt SD, Wang K (1995) Candidate gene for the chromosome 1 familial Alzheimer's disease locus. Science 269:973–977.

Mattis S (1983) Dementia rating scale. Odessa: Psychological Assessment Resources

Raux G, Gantier R, Martin C, Pothin Y, Brice A, Frebourg T, Campion D (2000) A novel presenilin 1 missense mutation (L153V) segregating with early-onset autosomal dominant Alzheimer's disease. Human Mutation 16:95.

Reitan R, Wolfson D (1985) The Halstead-Reitan neuropsychological test battery. Tuscon: Neuropsychological Press

Saura CA, Choi SY, Beglopoulos V, Malkani S, Zhang D, Shankaranarayana Rao BS, Chattarij S, Kelleher RJ 3rd, Kandel ER, Duff K, Kirkwood A, Shen J (2004) Loss of presenilin function causes impairments of memory and synaptic plasticity followed by age-dependent neurodegeneration. Neuron 42:23–36.

Scheuner D, Eckman C, Jensen M, Song X, Citron M, Suzuki N, Bird TD, Hardy J, Hutton M, Kukull W, Larson E, Levy-Lahad E, Viitanen M, Peskind E, Poorkaj P, Schellenberg G, Tanzi R, Wasco W, Lannfelt L, Selkoe D, Younkin S (1996) Aβ42(43) is increased *in vivo* by the PSEN1/2 and APP mutations linked to familial Alzheimer's disease. Nature Med 2:864–870.

Shen J, Bronson RT, Chen DF, Xia W, Selkoe DJ, Tonegawa S (1997) Skeletal and CNS defects in Presenilin-1-deficient mice. Cell 89:629–639.

Sherrington R, Rogaev EI, Liang Y, Rogaeva EA, Levesque G, Ikeda M, Chi H, Lin C, Li G, Holman K; Tsuda T, Mar L, Foncin JF, Bruni AC, Montesi MP, Sorbi S, Rainero I, Pinessi L, Nee L, Chumakov I, Pollen D, Brookes A, Sanseau P, Polinsky RJ, Wasco W, Da Silva HAR, Haines JL, Pericak-Vance MA, Tanzi RE, Roses AD, Fraser PE, Rommens JM, St George-Hyslop PH (1995) Cloning of a gene bearing missense mutations in early-onset familial Alzheimer's disease. Nature 375:754–760.

Tang-Wai D, Lewis P, Boeve B, Hutton M, Golde T, Baker M, Hardy J, Michels V, Ivnik R, Jack C, Petersen R (2002 Familial frontotemporal dementia associated with a novel presenilin-1 mutation. Dement Geriat Cogn Disord 14:13–21.

Wechsler D (1987) Weschler memory scale - revised. New York: Psychological Corporation.

Frontotemporal Dementias: Genotypes and Phenotypes

Virginia M.-Y. Lee[1]

Introduction

Frontotemporal dementias (FTDs) refer to a heterogeneous group of sporadic and familial neurodegenerative disorders in which the clinical manifestations are linked to pathologic processes that specifically target the frontal and temporal lobes, leading to relentlessly progressive and selective brain degeneration (Kertesz 1997; Neary et al. 1998). FTDs are characterized clinically by abnormal social conduct, loss of executive functions and personality changes (Kersetz 1997; McKhann et al. 2001; Neary et al. 1998; Lund and Manchester Groups 1994). However, the specific clinical manifestations of FTDs are determined by the topographic distribution of pathology affected by a neurodegenerative disease process; consequently, other diverse disorders – including Alzheimer's disease (AD), Lewy body disease, Huntington's disease, and prion disorders – may also present with frontotemporal lobe symptoms that closely resemble those seen in classic FTDs (Trojanowski and Lee 2003). Pick's disease (PiD) has long been considered the prototypical FTD. However, the discovery that family members with the same *tau* gene mutations that are pathogenic for FTDP-17 have diverse clinical phenotypes – including movement disorders such as progressive supranuclear palsy (PSP) and corticobasal degeneration (CBD), memory dysfunction similar to that observed in AD and argyrophilic grains disease (AGD), as well as the classic picture of FTD – suggested that all of these neurodegenerative disorders are part of the spectrum of FTD syndromes (Neary et al. 1998). Indeed, from a neuropathological perspective, the spectrum of FTDs includes those with and without abundant tau pathology (Fig. 1). About 30% of FTDs contain prominent tau abnormalities – characterized by the accumulation of insoluble hyperphophorylated tau proteins in neurons, glial cells or both neurons and glia – and, together with other neurodegenerative disorders with tau pathology, these FTDs are now classified as tauopathies. Another subtype of FTD is known as dementia lacking distinctive histopathology (DLDH), or frontotemporal lobar degeneration (FTLD; Knopman et al. 1990). Although DLDH brains do not contain tau lesions, DLDH is considered a tauopathy because there is a loss or marked reduction in the levels of brain tau proteins in DLDH compared to normal, age-matched controls (Zhukareva

1 The Center for Neurodegenerative Disease Research, Department of Pathology and Laboratory Medicine, and the Institute on Aging, the University of Pennsylvania School of Medicine, Philadelphia, PA, 19104-4283

Cummings et al.
Genotype – Protéotype – Phenotype
Relationships in Neurodegenerative Diseases

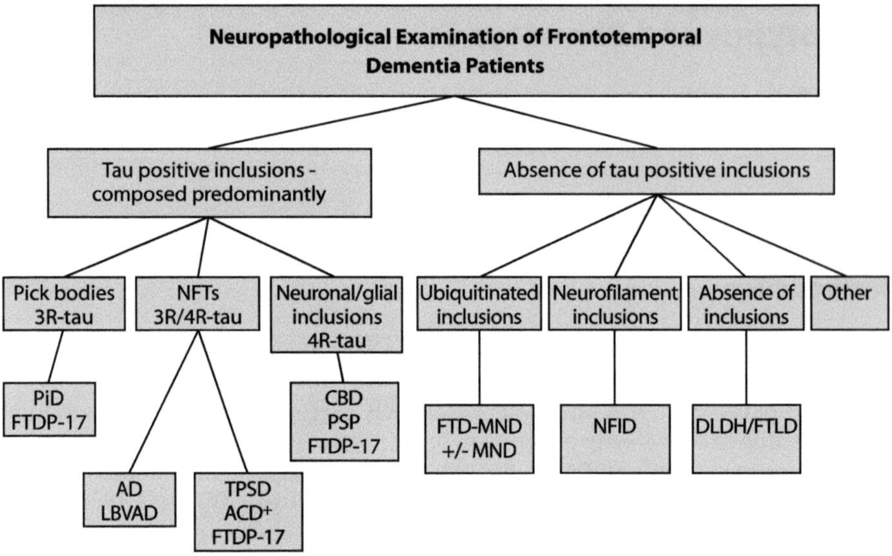

Fig. 1. Algorithm for the neuropathologic diagnosis of patients with FTD. This algorithm represents an oversimplification of the schema used to diagnose FTD patients and does not reflect the complexity of the cases or the commonly observed occurrence of multiple pathologies.

et al. 2001, 2003). A second major subtype of FTD is characterized by the presence of ubiquitin-positive and tau-, α-synuclein, and neurofilament (NF)-negative lesions similar to those observed in motor neuron disease (MND); this FTD variant is known as FTD-MND (see Fig. 1; Mitsuyama and Takamiya 1979; Neary et al. 1998). Finally, a recently described FTD subtype is characterized by the accumulation of intracellular inclusions that are negative for tau or α-synuclein but contain NF proteins and α-internexin; this variant is referred to as an FTD neurofilamentopathy, or neuronal intermediate filament inclusion disease (NIFID) (3,4). This review will summarize the current understanding of the genotypes and phenotypes of FTDs.

Genetics of FTD

The genetics of familial FTDs is complex and reflects broadly their sporadic counterparts as outlined above. Many patients who develop familial neurodegenerative disorders with a clinical phenotype consistent with PiD, PSP, CBD, and FTD turn out to have a *tau* gene mutation and therefore are part of FTDP-17 (McKhann et al. 2001). On the other hand, there are familial FTDs with similarly diverse clinical phenotypes and tau pathologies but without identifiable *tau* gene mutations (Lendon et al. 1998; Morris et al. 1984). Moreover, a large FTD-MND kindred with linkage to chromosome 17 developed abundant tau and α-synuclein pathology but was without *tau* as well as α-synuclein mutations (Wilhelmsen et al. 2004). Thus, FTD-associated genes other than tau that cause tau pathologies remain to be identified. Furthermore, other FTD kindred with linkage to chromosome 17 lacked both tau pathologies and *tau* gene mutations, indicating that other, related genes on chromosome 17 also cause FTD in an autosomal-dominant manner. Loci on other chromosomes have also been implicated in FTD. For instance, families with inclusion body myopathy, Paget's disease, and co-incidental DLDH were linked to chromosome 9 (Waggoner et al. 2002). In addition, other FTD kindreds with FTD-MND pathology showed strong linkage to chromosome 3 (Brown et al. 1995), and a distinct locus on chromosome 9 (Hosler et al. 2000). Thus, additional gene and risk factors for this group of heterogeneous familial FTDs remain to be identified.

Clinical features of FTD

The clinical phenotype of patients with both familial and sporadic FTDs also exhibits a wide range of specific cognitive and behavioral deficits compared with the genotype and the neuropathological phenotype. For example, clinical characteristics of FTD patients include impairment of executive functioning that interferes with strategic processing and working memory, difficulty with initiative and attention, non-fluent speech that progresses to mutism, and a bizarre affect that includes disinhibited, hypersexual and hyperoral behaviors (Foster et al. 1997; Kertesz 1997; Knopman et al. 1990; Lieberman et al. 1998; Morris et al. 1984; Neary 1994; Neary et al. 1998; Lund and Manchester Groups 1994; Turner et al. 1996). In addition, FTD patients often develop extra-pyramidal signs such as rigidity and gait disturbance that may be accompanied by a resting tremor. Other features of sporadic and familial FTDs include 1) an inattentive state with poor initiative, 2) a dysexecutive syndrome, 3) a progressive non-fluent aphasia, and 4) a disinhibited bizzare affect. However, the latter clinical features of FTD are not well characterized but are important for understanding FTDs for several reasons: 1) Many of the clinical symptoms in FTD patients overlap with those in other dementing diseases, most notably AD. In fact, the prevalence of FTDs was underestimated in the past because they were confused with AD, but increasing awareness suggests that they may constitute up to 20% of neurodegenerative dementias; 2) Given the heterogeneity of FTDs, it is important to define subgroups of these FTD patients based on cognitive and behavioral evaluations. 3) Once identified, clinical subgroups of FTDs can be correlated with the neuropathological variants of FTDs (see Fig. 1). 4) Emerging genetic and biochemical analyses of FTDP-17 provide support for the hypothesis that the FTDP-17 genotype predicts the abnormal biochemical tau phenotype, but it is unclear whether or not it also correlates with clinical phenotype. 5) Differentiation of FTDs from other dementing diseases on clinical grounds will facilitate efforts to dissect out the underlying mechanisms of the neurobehavioral deficits specific to each disorder, as well as to conduct cost-effective and informative clinical trials of new therapeutic agents that are likely to emerge from the rapidly accruing insights into the basic mechanisms of different neurodegenerative diseases. Thus, increased understanding of the clinical phenotypes of FTDs has several important implications for FTDs as well as for AD and other neurodegenerative disorders.

FTD Tauopathies

One of the major neuropathologies found in FTD patients is filamentous tau inclusions and related pathologies that accumulate in neurons and/or glia (Lee et al. 2001). To begin to understand how tau pathology causes neuronal and perhaps glial degeneration, it is important to consider the normal biology of tau proteins. Tau belongs to the family of MT-associated proteins, and tau proteins bind to MTs and regulate MT functioning. In the adult human brain, alternative mRNA splicing of 11 exons in the *tau* gene generates 6 tau isoforms (Goedert et al. 1989). Alternative splicing of exon (E) 2 and E3 of the tau gene generates isoforms with or without a 29 or 58 amino acid-long amino-terminal insert, and alternative splicing of E10 gives rise to three tau isoforms with three MT binding repeats (3R-tau) or three tau isoforms with 4 MT binding repeats (4R-tau). These repeats are nearly identical, 31 amino acid-long motifs (Goedert et al. 1989). While much remains to be learned about how the alternative splicing of the tau gene is regulated, the demonstration that the ratio of 3R-tau to 4R-tau in the normal brain is ~1 (Hong et al. 1998) supports the idea that this splicing is tightly regulated.

In 1994, the first FTDP-17 kindred was linked to chromosome 17q21.22, the same region containing the *tau* gene (Wilhelmsen et al. 1994). Subsequently, in June 1998, Schellenberg and co-workers reported the first pathogenic *tau* gene mutation in a FTDP-17 kindred (Seattle family A; Poorkaj et al. 1998). Shortly thereafter, a large number of intronic and exonic tau gene mutations in multiple FTDP-17 kindreds were reported in several studies, confirming that the FTDP-17 locus is indeed the *tau* gene (Hutton et al. 1998; Iijima et al. 1999; Spillantini et al. 1998). The extent and the variety of the different *tau* gene mutations discovered so far directly implicate *tau* dysfunction as a major cause of familial FTDP-17. These initial observations emphasize that the fine regulation of tau isoform expression in neurons and glial cells is exquisitely sensitive to perturbation and that multiple mechanisms of disease pathogenesis can be initiated by the different *tau* gene mutations. To date, >30 different *tau* gene mutations have been identified (Table 1), and they cause FTDP-17 by reducing the binding of tau to MTs, increasing the propensity of tau to assemble into fibrils and/or by altering E10 splicing and consequently altering the ratio of 3R- to 4R-tau (Lee et al. 2001). Recent studies from several laboratories have demonstrated that multiple mechanisms, including regulatory elements in introns as well as missense mutations within the coding region, are operative in regulating E10 splicing. This increase in E10 splicing augments the expression of 4R-tau proteins in the brains of FTDP-17 patients with these specific mutations.

Significantly, the discovery of E10 splice mutations in a subset of FTDP-17 kindreds provided important clues for developing a better understanding of seemingly sporadic FTDs, including those assigned the postmortem neuropathological diagnosis of PiD, CBD, PSG or PSP. The majority of the *tau* gene mutations are located within or near the MT binding repeat regions and are inherited in an au-

Table 1. Tau mutations in FTDP-17a

Mutation	Location	Exon 10 splicing	MT binding	Pheno-type	Reference
R5H	Exon 1	No change	Reduced	FTDP-17	Josephs et al. 2003
R5L	Exon 1	No change	Reduced	PSP-like	Kertesz 1997
K257T	E9, R1	No change	Reduced	PiD-like	Kinoshita et al. 1997; Knopman et al. 1990
I260V	E9, R1	ND	ND	NA	Lariviere and Julien 2004
L266V	E9, R1	Decreasedb	Reduced	PiD-like	Lee et al. 2001
G272V	E9, R1	No change	Reduced	FTDP-17	Leigh et al. 1988
E9+33	I9	ND	NA	NA	Lendon et al. 1998
N279K	E10, IR1-2	Increasedc	Variable	PSP-like	Lieberman et al. 1998
ΔK280	E10, IR1-2	Decreased	Reduced	FTDP-17	Lendon et al. 1998
L284L	E10, IR1-2	Increased	NA	AD-like	Lowe et al. 1988
N296N	E10, R2	Increased	NA	CBD-like	Mann 1998
N296H	E10, R2	Increased	Decreased	FTDP-17	Mann et al. 1993
ΔN296	E10, R2	No change	Decreased	PSP-like	McKhann et al. 2001
P301L	E10, R2	No change	Reduced	FTDP-17	Leigh et al. 1988
P301S	E10, R2	No change	Reduced	CBD-like, FTDP-17	Meyer 1929; Mitsuyama 2000
S305N	E10, IR2-3	Increased	No effect	CBD-like	Mitsuyama and Takamiya 1979; Morris et al. 1984
S305S	E10, IR2-3	Increased	NA	PSP-like	Nakano 2000
E10+3	I10	Increased	NA	FTDP-17	Neary et al. 1988
E10+11	I10	Increased	NA	FTDP-17	Neary 1994
E10+12	I10	Increased	NA	FTDP-17	Neary et al. 1998
E10+13	I10	Increased	NA	NA	Leigh et al. 1988
E10+14	I10	Increased	NA	PSP-like, FTDP-17	Leigh et al. 1988
E10+16	I10	Increased	NA	AD-, PiD-, PSP-, CBD-like, FTDP-17	Leigh et al. 1988; Nicholl et al. 2003
L315R	E11	No change	Reduced	PiD-like	Okamoto et al. 1991
S320F	E11	No change	Reduced	PiD-like	Poorkaj et al. 1998
Q336R	E12	No change	Increased	PiD-like	Poorkaj et al. 2002

Table 1. *Continued*

Mutation	Location	Exon 10 splicing	MT binding	Pheno-type	Reference
V337M	E12, IR3-4	No change	Reduced	FTDP-17	Spillantini et al. 1998
E342V	E12, IR3-4	Increased	ND	FTDP-17	Talbot 1996
S352L	E12, IR3-4	ND	No effect	Atypical	The Lund and Manchester Groups 1994
K369I	E12, IR3-4	ND	Reduced	PiD-like	Trojanowski and Lee 2003
G389R	E13	No change	Reduced	PiD-like	Turner et al. 1996
R406W	E13	No change	Reduced	PSP-like	Leigh et al. 1988

[a] Abbreviations: E = exon; I = intron; R = MT binding repeat; IR = inter-repeat regions; ND = not determined; NA = not applicable [b]Decreased indicates reduced exon 10 utilization. [c]Increased indicates enhanced exon 10 utilization.

tosomal-dominant manner. However, a number of "unusual" *tau* gene mutations were discovered. First, an N-terminal exon 1 *tau* mutation, R5L (Poorkaj et al. 2002), that is pathogenic for FTD was identified, and a R5H *tau* mutation also was identified in a Japanese PSP patient (Hayashi et al. 2002). Notably, recombinant R5L tau fibrillizes more readily than wild type (WT) tau (Gamblin et al. 2003). Second, an autosomal recessive *tau* gene mutation, S352L, was identified in an unusual kindred. The affected individuals of this pedigree presented with respiratory failure but subsequently developed a mild PSP-like phenotype (Nicholl et al. 2003). Third, incomplete penetrance was observed in a Dutch family with the L315R *tau* mutation (van Herpen et al. 2003). Thus, the genetics of *tau* mutations in FTD tauopathies are complex and the identification of additional *tau* gene mutations or novel genes with mutations will be crucial for increasing understanding of the etiology and pathogenesis of this complex disorder.

Studies conducted in the last few years on tau pathologies in different FTDs and AD provide important insights into the similarities and differences between these FTD tauopathy variants and AD tau pathologies. For example, the density of glial tau pathologies, particularly in white matter, is found in FTD and not in AD, which distinguishes AD from FTD tau pathologies. Moreover, the regional distribution of tau pathologies in AD differs from that of FTD tauopathies, wherein subcortical and brainstem tau lesions are prominent features. Finally, although elevated levels of CSF tau are relatively specific AD biomarkers, FTDs do not show consistent increases in CSF tau (Clark et al. 2003). Taken together, these and other findings support the hypothesis that distinct pathogenic mechanisms cause FTD tauopathies compared to the tau pathologies in AD.

FTD with MND-type inclusions

In 1929, Meyer first noted an association between MND and dementia (Meyer, 1929). Since that time, this combination of clinical signs and symptoms has been increasingly recognized (Mitsuyama and Takamiya 1979; Neary et al. 1988)(Mitsuyama and Takamiya, 1979; Neary et al., 1990). Patients typically present with FTD and neurological signs of MND appear later. However, occasionally the dementia and physical symptoms present concurrently (Mitsuyama 2000; Nakano 2000; Talbot 1996). In some instances, FTD occurs in the absence of clinical signs of MND, despite the presence of the characteristic pathological changes in the brain (Jackson et al. 1996). Currently, FTLD-MND, also referred to as "motor neuron disease-inclusion dementia," is one of the more common neuropathological diagnoses in patients with FTD. However, the relationship between FTD patients with and without MND, as well as MND alone, remains unclear.

The gross pathological changes of FTD are relatively similar in FTD, regardless of the underlying histopathology (Mann 1998; The Lund and Manchester Groups 1994). In addition, the anterior roots of the spinal cord are often atrophic in patients with prominent MND (Mitsuyama and Takamiya 1979; Neary et al. 1988). Loss of motor neurons in the spinal cord and brainstem is a feature of FTD-MND that is similar to that observed in classic MND, but the changes in motor neurons are generally more prominent in those patients with clinically advanced MND (Jackson et al. 1996). Histochemical and immunohistochemical analyses reveal no senile plaques, amyloid angiopathy, tau-immunoreactive neurofibrillary lesions, or α-synuclein pathology in FTLD-MND. In contrast, the defining histological feature is the presence of cytoplasmic neuronal inclusions in the granule cells of the dentate gyrus and the superficial layers of the neocortex, most prominently in the external granular layer (Jackson et al. 1996; Okamoto et al. 1991; Wightman et al. 1992). These inclusions are similar morphologically to Pick bodies, except they are immunoreactive with antibodies to ubiquitin but non-reactive with antibodies for tau and are not detected with conventional silver stains. Ubiquitin-immunoreactive dystrophic neurites are also detected in affected neocortices. Ultrastructurally, the inclusions are composed of filaments 10 to 15 nm in diameter (Iseki et al. 2001; Kinoshita et al. 1997). In addition, there are often ubiquitinated inclusions within motor neurons of the anterior horns of the spinal cord and brainstem that are identical to those observed in classic MND (Fig. 1; Leigh et al. 1988; Lowe et al. 1988).

Relationship between FTLD-MND and classic MND

While the relationship between FTLD-MND and classic MND remains uncertain, there are numerous similarities between the two disorders. In classic MND, there is often frontal lobe dysfunction in later stages of the illness, despite the absence of cortical or hippocampal pathology. The ubiquitinated inclusions in the brain and spinal cord of FTLD-MND patients are similar to those identified in the spinal cord of classic MND patients (Jackson et al. 1996; Okamoto et al. 1991 Wightman et al. 1992). Furthermore, several families with FTD-MND have been described in which affected individuals clinically manifest FTD alone, MND alone, or both (Gunnarsson et al. 1991; Jackson et al. 1996; Mann et al. 1993). Thus, it is possible that FTLD-MND and classic MND represent different patterns of expression of a common disease process with an overlapping topographic distribution of pathology. The identification of mutations and/or polymorphisms in kindreds with FTLD-MND will provide insight into the pathogenesis of the disorder. However, it remains unclear whether FTLD-MND represents a primary disorder in ubiquitination or pathology of a distinct, as of yet unidentified, protein or proteins that are ubiquitinated during the process of aggregation.

Frontotemporal lobar degeneration

In 1974, Constantinidis and colleagues (1974) subclassified PiD into several categories depending on the presence or absence of Pick bodies and/or ballooned neurons. More recent neuropathological criteria for the classification of FTD require Pick bodies for the diagnosis of Pick's disease (McKhann et al. 2001; The Lund and Manchester Groups 1994). A subset of these FTD patients (Constantinidis Group C) lacks filamentous inclusions within either neurons or glia. This subset, which accounts for up to 30% of patients with FTD, has received a variety of names, including dementia of frontal lobe type, frontal lobe degeneration of non-Alzheimer type, and dementia lacking distinctive histopathology (Brun 1987; Knopman et al. 1990; Mearu et al. 1988). A recent report from the Work Group on FTD and PiD settled upon the name "frontal temporal lobar degeneration" (FTLD; McKhann et al. 2001).

Although the frontal and temporal lobe atrophy and variable depigmentation of the substantia nigra are the same as are observed for FTLD-MND described above (except for the notable absence of atrophy of the anterior roots of the spinal cord and variable degeneration of the hippocampal formation, basal ganglia, and substantia nigra consisting of neuron loss and gliosis), immunohistochemistry (IHC) reveals no (or limited) filamentous inclusions identified with antibodies specific for amyloid, tau, α-synuclein, or ubiquitin, and there is no protease-resistant prion protein (McKhann et al. 2001; The Lund and Manchester Groups 1994). However, Zhukareva et al. (2001, 2003) identified the first unique feature in FTLD patients: in a subset of these patients, they noted the selective loss of tau protein expression but relative preservation of tau mRNA levels. This finding suggests that the level of tau protein is controlled post-transcriptionally. Moreover, the loss of functional tau expression may disrupt axonal transport, leading to neurodegeneration of affected neurons.

FTD with neuronal intermediate filament inclusion disease (NIFID)

A subtype of FTD has been shown to develop neuronal cytoplasmic inclusions comprised of neuronal intermediate filaments that are tau- and α-synuclein-negative but variably ubiquitinated; it is designated as neuronal intermediate filament inclusion disease (NIFID). Although all neuronal intermediate filament proteins, including NF triplet protein and α-internexin, are present in the inclusions, α-internexin is more abundant and hence has been identified as a major component of the pathological inclusions of this subtype of FTD (Cairns et al. 2004a,b). There are six types of IF proteins classified by gene structure and sequence homology. The term "intermediate" derives from their diameter (10–12 nm), which is intermediate between microtubules (25 nm) and microfilaments (7–10 nm). Five major neuronal IF proteins are expressed in the adult human CNS: three NF proteins – light (NF-L), medium (NF-M), and heavy (NF-H) subunits of approximately 68 kDa, 145 kDa, and 200 kDa, respectively; peripherin, of 57kDa; and α-internexin, of 66kDa. NFs and α-internexin genes have homologous intron-exon organization and are type IV, whereas peripherin encodes a type III IF protein resembling vimentin (Lariviere and Julien 2004). The IF proteins have a tripartite structure: a central rod domain of about 300 amino acids formed from a highly conserved α-helix, and amino- and carboxy-terminal regions called head and tail domains, respectively, which are less conserved (Lariviere and Julien 2004). NF inclusions are commonly found in tangle-bearing neurons in AD and in Lewy body-bearing neurons in Parkinson's disease (PD) and exist as spheroids in amyotrophic lateral sclerosis (ALS). However, the presence of α-internexin distinguishes NIFID from other diseases with intermediate filament inclusions since α-internexin is rarely detected in neurodegenedarative diseases. The accumulation of abnormal neuronal intermediate filament aggregates in neurons might be a consequence of dysregulation of protein synthesis, failure of axonal transport, abnormal phosphorylation and proteolysis.

Conclusions

The complexity of FTDs is beginning to be elucidated. Diverse clinical phenotypes including behavior abnormalities, aphasia, parkinsonism, motor neuron degeneration and dementia are part of the spectrum of FTD syndrome. Although the molecular mechanism that accounts for this clinical heterogeneity is unclear, the overlapping clinical phenotypes suggest an inter-dependent relationship in the manifestation of these diverse phenotypes. Investigations into the genetics of FTDs may facilitate our understanding of the etiology and pathogenesis of FTDs, and it is evident that autosomal-dominant and autosomal-recessive inheritance in multiple genes as well as complex traits will be implicated in FTDs. Although *tau* gene mutations on chromosome 17 are found to be causal in some families with FTDP-17 and tau polymorphisms are risk factors for FTD, they are responsible for only a fraction of familial FTDs. Nevertheless, the accumulation of tau lesions in patients with known FTDP-17 mutations supports the concept that the aggregation of mutant brain protein can lead directly to neurodegeneration. Since the majority of FTDs do not accumulate tau lesions and since a sizable percentage of FTDs develop ubiquitin-positive and tau- and α-synuclein-negative inclusions, the identification of the protein entities that are ubiquitinated represents a major challenge in the field, and their identification will provide additional insights into disease pathogenesis as well as into the development of novel strategies for treatment and prevention.

Acknowledgments

Support for this work was provided by grants from the National Institute on Aging of the National Institutes of Health. VM-YL is the John H. Ware, 3rd Professor of Alzheimer's Research. We would also like to thank the members of the Center for Neurodegenerative Disease Research who contributed to the work, and the many patients studied and their families, for making the research reviewed here possible.

References

Brown JM, Ashworth A, Gydesen S Sorensen A, Rossor M, Hardy J, Collinge J (1995) Familial non-specific dementia maps to chromosome 3. Human Mol Genet 4:1625–1628.

Brun A (1987) Frontal lobe degeneration of non-Alzheimer type. I. Neuropathology. Arch Gerontol Geriatr 6:193–208.

Cairns NJ, Jaros E, Perry RH, Armstrong RA (2004a) Temporal lobe pathology of human patients with neurofilament inclusion disease. Neurosci Lett 354: 245–247.

Cairns NJ, Zhukareva V, Uryu K, Zhang B, Bigio E, Mackenzie IR, Gearing M, Duyckaerts C, Yokoo H, Nakazato Y, Jaros E, Perry RH, Lee VM, Trojanowski JQ. (2004) α-Internexin is present in the pathological inclusions of neuronal intermediate filament inclusion disease. Am J Pathol 164:2153–2161.

Clark CM, Ewbank D, Peskind E, Galasko D, Morris JC, Farlow M, Quinn J, Kaye ., Knopmna D, Arai H, Doody RS, DeCarli C, Leight S, Lee VM-Y, Trojanowski JQ (2003) CSF tau and β-amyloid: How well do these biomarkers reflect autopsy confirmed dementia diagnoses? Arch Neurol 60:1696–1702.

Constantinidis J, Richard J, Tissot R (1974) Pick's disease. Histological and clinical correlations. Eur Neurol 11: 208–217.

Foster NL, Wilhelmsen KC, Sima AA, Jones MZ, D'Amato CJ, Gilman S (1997) Frontotemporal dementia and parkinsonism linked to chromosome 17: a consensus statement. Ann Neurol 41:706–715

Gamblin TC, Berry RW, Binder LI (2003) Tau polymerization: role of the amino terminus. Biochemistry 42:2252–2257.

Goedert M, Spillantini MG, Jakes R, Rutherford D, Crowther RA (1989) Multiple isoforms of human microtubule-associated protein tau: sequences and localization in neurofibrillary tangles of Alzheimer's disease. Neuron 3:519–526.

Gunnarsson LG, Dahlbom K, Strandman E (1991) Motor neuron disease and dementia reported among 13 members of a single family. Acta Neurol Scand 84: 429–433.

Hayashi S, Toyoshima Y, Hasegawa M, Umeda Y, Wakabayashi K, Tokiguchi S, Iwatsubo T, Takahashi H (2002) Late-onset frontotemporal dementia with a novel exon 1 (Arg5His) tau gene mutation. Ann Neurol 51:525–530

Hong M, Zhukareva V, Vogelsberg-Ragaglia V, Wszolek Z, Reed L, Miller BI, Geschwind DH, Bird TD, McKeel D, Goate A, Morris JC, Wilhelmsen KC, Schellenberg GD, Trojanowski JQ, Lee VM-Y (1998) Mutation specific functional impairments in distinct tau isoforms of FTDP-17. Science 282:1914–1917.

Hosler BA, Siddique T, Sapp PC, Sailor W, Huang MC, Hossain A, Daube JR, Nance M, Fan C, Kaplan J, Hung WY, McKenna-Yasek D, Haines JL, Pericak-Vance MA (2000) Linkage of familial amyotrophic lateral sclerosis with frontotemporal dementia to chromosome 9q21-q22. JAMA 284:1664–1669.

Hutton M, Lendon CL, Rizzu P, Baker M, Froelich S, Houlden H, Pickering-Brown S, Chakraverty S, Isaacs A, Grover A, Hackett J, Adamson J, Lincoln S, Dickson D, Davies P, Petersen R, Stevens M, De Graaff E, Wauter Erwin, Van Baren J, Hillebrand M, Joosse M, Kwon JM, Nowotny P, Che LK, Norton J, Morris JC, Reed LA, Trojanowski J, Basun H, Lannfelt L, Neystat M, Fahn S, Dark F, Tannenberg T, Dodd PR, Hayward N, Kwok JBJ, Schofield PR, Andreadis A, Snowden J, Craufurd D, Neary D, Owen F, Oostra BA, Hardy J, Goate A, Van Swieten J, Mann D, Lynch T, Heutink P (1998) Association of missense and 5'-splice- site-mutations in tau with the inherited dementia FTDP-17. Nature 393:702–705.

Iijima, M, Tabira T, Poorkaj P, Schellenberg GD, Trojanowski JQ, Lee VM-Y, Schmidt ML, Takahashi K, Nabika T, Matsumoto T, Yamashito Y, Yoshioka S, Ishino H (1999) A familial presenile dementia with abundant tau positive neuronal and glial tangles linked to a mutation in exon 10 of the tau gene. NeuroReport 10:497–501.

Iseki E, Matsumura T, Marui W, Hino H, Odawara T, Sugiyama N, Suzuki K, Sawada H, Arai T, Kosaka K (2001) Familial frontotemporal dementia and parkinsonism with a novel N296H mutation in exon 10 of the tau gene and a widespread tau accumulation in glial cells. Acta Neuropathol 102:285–292.

Jackson M, Lennox G, Lowe J (1996) Motor neurone disease-inclusion dementia. Neurodegeneration 5: 339–350.

Josephs KA, Holton JL, Rossor MN, Braendgaard H, Ozawa T, Fox NC, Petersen RC, Pearl GS, Ganguly M, Rosa P, Laursen H, Parisi JE, Waldemar G, Quinn NP, Dickson DW, Revesz T (2003) Neurofilament inclusion body disease: a new proteinopathy? Brain 126:2291–2303.

Kertesz A (1997) Frontotemporal dementia, Pick's disease, and corticobasal degeneration. Arch Neurol 54:1427–1428.

Kinoshita A, Tomimoto H, Suenaga T, Akiguchi I, Kimura J (1997) Ubiquitin-related cytoskeletal abnormality in frontotemporal dementia: immunohistochemical and immunoelectron microscope studies. Acta Neuropathol (Berl), 94: 67–72.

Knopman DS, Mastri AR, Frey II WH, Sung JH, Rustan T (1990) Dementia lacking distinctive histologic features: A common non-Alzheimer degenerative dementia. Neurology 40: 251-256.

Lariviere RC, Julien JP (2004) Functions of intermediate filaments in neuronal development and disease. J Neurobiol 58:131–148.

Lee VM-Y, Goedert M, Trojanowski JQ (2001) Neurodegenerative tauopathies. Ann Rev Neurosci 24: 1121–159.

Leigh, PN, Anderton BH, Dodson A, Gallo JM, Swash M, Power DM (1988) Ubiquitin deposits in anterior horn cells in motor neurone disease. Neurosci Lett 93:197–203.

Lendon CL, Lynch T, Norton J, McKeel Jr DW, Busfield F, Craddock N, Chakraverty S, Gopalakrishnan G, Shears SD, Grimmett W, Wilhelmsen KC, Hansen L, Morris JC, Goate AM (1998) Herediary dysphasic disinhibition dementia: A frontal dementia linked to 17q21-22. Neurology 50: 1546–1555.

Lieberman AP, Trojanowski JQ, Lee VM-Y, Balin BJ, Ding X-S, Grossman M (1998) Cognitive, neuroimaging, and pathological studies in a patient with Pick's disease. Ann Neurol 43:259-265.

Lowe J, Lennox G, Jefferson D, Morrell K, McQuire D, Gray T., Landon M, Doherty FJ, Mayer RJ (1988) A filamentous inclusion body within anterior horn neurones in motor neurone disease defined by immunocytochemical localization of ubiquitin. Neurosci Lett 94:203–210.

Mann DM (1998). Dementia of frontal type and dementias with subcortical gliosis. Brain Pathol 8:325–338.

Mann DM, South PW, Snowden JS, Neary D (1993) Dementia of frontal lobe type: neuropathology and immunohistochemistry. J Neurol Neurosurg Psychiat 56:605–614.

McKhann GM, Albert MS, Grossman M, Miller B, Dickson DW, Trojanowski JQ (2001) Clinical and pathological diagnosis of frontotemporal dementia: report of Work Group on Frontotemporal Dementia and Pick's Disease. Arch Neurol 58:1803–1809.

Meyer A (1929). Über eine der amyotrophischen Lateralsklerose nahestende Erkrankung mit psychischen Storungen. Z Ges Neurol Psychiat 121:107–128.
Mitsuyama Y (2000) Dementia with motor neuron disease. Neuropathology, 20 (Suppl): S79–S81.
Mitsuyama Y, Takamiya S (1979). Presenile dementia with motor neuron disease in Japan. A new entity? Arch Neurol 36:592–593.
Morris JC, Monroe C, Banker BQ, Wright D (1984) Hereditary dysphasic dementia and the Pick-Alzheimer spectrum. Ann Neurol 16: 455–466.
Nakano I (2000) Frontotemporal dementia with motor neuron disease (amyotrophic lateral sclerosis with dementia). Neuropatholog, 20:68–75.
Neary D (1994) Classification of the dementias. Rev Clin Gerontol 4: 131–140.
Neary D, Snowden JS, Northen B, Goulding P (1988) Dementia of frontal lobe type. J Neurol Neurosurg Psychiat 51: 353–361.
Neary D, Snowdon JS, Gustafson L, Passant U, Stuss D, Black S, Freedman M, Kertesz A, Robert PH, Albert M, Boone K, Miller BL, Cummings J, Benson DF (1998) Frontotemporal lobar degeneration: A consensus on clinical diagnostic criteria. Neurology 51: 1546–1554
Nicholl DJ, Greenstone MA, Clarke CE, Rizzu P, Crooks D, Crowe A, Trojanowski JQ, Lee VM-Y, Heutink P (2003) An English kindred with a recessively inherited tauopathy due to a novel S352L mutation presenting with acute respiratory failure. Ann Neurol 54:682–686.
Okamoto K, Hirai S, Yamazaki T, Sun, XY, Nakazato Y (1991) New ubiquitin-positive intraneuronal inclusions in the extra-motor cortices in patients with amyotrophic lateral sclerosis. Neurosci Lett 129: 233–236.
Poorkaj P, Bird TD, Wijisman E, Nemens E, Garruto RM, Anderson L, Andreadis A, Wiederholt WC, Raskind M, Schellenberg GD (1998) Tau is a candidate gene for chromosome 17 frontotemporal dementia. Ann Neurol 43: 8150–825.
Poorkaj P, Muma NA, Zhukareva V, Cochran,EJ, Shannon KM, Hurtig H, Koller WC, Bird, Trojanowski J., Lee VM-Y, Schellenberg GD (2002) An R5L tau mutation in a subject with a progressive supranuclear palsy phenotype. Ann Neurol 52: 511–516.
Spillantini ML, Murrell JR, Goedert M, Farlow MR, Klud A, Ghetti B (1998) Mutation in the tau gene in familial multiple system tauopathy with presenile dementia. Proc Natl Acad Sci USA 95: 7737–7741.
Talbot PR (1996) Frontal lobe dementia and motor neuron disease. J Neural Transm Supp, 47: 125–132.
The Lund and Manchester Groups (1994) Clinical and neuropathological criteria for frontotemporal dementia. Neurol Neurosurg Psychiat 57: 416–418.
Trojanowski JQ, Lee VM-Y (2003) Frontotemporal dementias. In: Rosenberg RN, Prusiner SB, Di Mauro S, Barchi RL, Nestler EJ (eds) The molecular and genetic basis of neurological and psychiatric disease. 3rd Edition. Butterworth Heineman Press, Woburn, MA, pp. 343–350.
Turner RS, Kenyon LC, Trojanowski JQ, Gonatas N, Grossman M (1996) Clinical, neuroimaging, and pathologic features of progressive non-fluent aphasia. Ann Neurol 39:166–173.
van Herpen E, Rosso SM, Serverijnen LA, Yoshida H, Breedveld G, van de Graaf R, Kamphorst W, Ravid R, Willemsen R, Dooijes D, Majoor-Krakauer D, Kros JM, Crowther RA, Goedert M, Heutink P, van Swieten JC (2003) Variable phenotypic expression and extensive tau pathology in two families with the novel tau mutation L315R. Ann Neurol 54: 573–81.
Waggoner B, Kovach MJ, Winkelman M, Cai D, Khardori R, Gelber D, Kimonis VE (2002) Heterogeneity in familial dominant Paget disease of bone and muscular dystrophy. Am J Med Genet. 108: 187–91.
Wightman G, Anderson VE, Martin J, Swash M, Anderton BH, Neary D, Mann D, Luthert P, Leigh PN (1992). Hippocampal and neocortical ubiquitin-immunoreactive inclusions in amyotrophic lateral sclerosis with dementia. Neurosci Lett 139: 269–274.
Wilhelmsen KC, Lynch T., Pavlou E, Higgins M, Nygaard TG (1994) Localization of dishibition-dementia-Parkinsonism-amyotrophy complex to 17q21-22. Am J Genet 55: 1159–1165.

Wilhelmsen K, Forman MS., Rosen, H, Alving IA., Goldman J, Feiger J, Lee JV, Segall SK, Kramer JH, Lomen-Hoerth C, Rankin KP, Johnson J, Feiler HS, Weiner MW, Lee VM, Trojanowski JQ, Miller BL (2004) 17q-linked frontotemporal dementia-amylotrophic lateral sclerosis without tau mutation with tau and α-syn inclusions. Arch Neurol 61: 398–406.

Zhukareva V, Vogelsberg-Ragaglia V, Van Deerlin VMD, Bruce J, Shuck T, Grossman M, Clark CM, Arnold SE, Masliah E, Galasko D, Trojanowski JQ, Lee VM (2001) Loss of brain tau defines novel sporadic and familial tauopathies with frontotemporal dementia. Ann Neurol 49:165–175.

Zhukareva V, Sundarraj S. Mann D, Sjogren M, Clark C (2003) Selective reduction of soluble tau proteins in sporadic and familial FTDs: an international follow-up study. Acta Neuropathol 105:469–476.

Chromosome 17-linked Frontotemporal dementia with Ubiquitin-Positive, Tau-Negative Inclusions

Rosa Rademakers[1], *Julie van der Zee*[1], *Samir Kumar-Singh*[1], *Bart Dermaut*[1], *Marc Cruts*[1] *and Christine Van Broeckhoven*[1]

Summary

Familial forms of frontotemporal dementia (FTD) are in 10-43% of patients, caused by mutations in the gene encoding the microtubule associated protein tau (*MAPT*) located at chromosome 17q21. Neuropathologically, these patients are characterized by tau-positive depositions in brain. However, autosomal dominant forms of FTD without *MAPT* mutations have been reported, suggesting other tauopathy-related genetic defects. One such form is FTD linked to 17q21, with tau-negative but ubiquitine-positive neuronal inclusions or FTD-U. We previously reduced the candidate chromosomal region to 4.8 cM in a Dutch FTD-U family, 1083. A mutation in *MAPT* was excluded by genomic sequencing. More recently, we identified three Belgian FTD families of which two, DR2 and DR8, showed linkage to the 17q21 region. Both families shared a common haplotype in an 8.04 cM region, indicating that they are genetically related to a common founder. In the third family, DR7, we obtained an autopsy confirmation of the characteristic ubiquitin-positive, tau-negative neuronal inclusions. Currently, there are 11 FTD families linked to 17q21 that do not segregate a *MAPT* mutation, of which five are conclusively linked (LOD score > 3). Together the data suggest that FTD-U could represent an important subtype of FTD, and that identification of the underlying gene defect might significantly contribute to our understanding of the pathomechanism leading to neurodegeneration in this dementia subtype.

Introduction

Frontotemporal dementia (FTD) is an important neurodegenerative disease representing approximately 5% of all dementia patients and 10–20% of patients with an onset age below 65 years (Neary et al. 1998; Ratnavalli et al. 2002). Clinically, FTD patients present with personality changes and disinhibited behavior, often accompanied by a gradual and progressive language dysfunction (McKhann et al. 2001). A definite diagnosis of FTD requires neuropathological brain examination. Unlike other forms of neurodegenerative disorders such as Alzheimer's disease (AD) or Creutzfeldt-Jacob's disease, FTD encompasses considerable pathological

[1] Department of Molecular Genetics, Flanders Interuniversity Institute for Biotechnology (VIB), University of Antwerp (UA), Antwerpen, Belgium

Cummings et al.
Genotype – Protéotype – Phenotype
Relationships in Neurodegenerative Diseases

heterogeneity. Three broad subdivisions have been recognized depending on the profile of immunohistochemical staining and the pattern of intracellular inclusions (Morris et al. 2001; Hodges et al. 2003). The first group contains FTD patients with tau-positive pathology, including patients with Pick bodies and those with diffuse tau-positive neuronal and astrocytic immunoreactivity. The second class consists of patients with a variable number of tau-negative, ubiquitin-positive intraneuronal cytoplasmic inclusions in frontotemporal cortex or the dentate gyrus (FTD-U), whereas the third class of patients is characterized by dementia lacking distinctive histopathology (DLDH). Many studies have recently attempted to estimate the prevalence of each of these three pathological FTD subtypes in extended populations (Hodges et al. 2004; Paviour et al. 2004; Josephs et al. 2004; Lipton et al. 2004). They showed that, in general, only a third of all FTD patients are tau-positive; conversely, the relative contribution of FTD-U and DLDH significantly varied between different studies and is still debated.

A positive family history of dementia is found in approximately 40% of FTD patients and, in the majority of these patients, the disease is inherited as an autosomal dominant trait (Stevens et al. 1998; Rosso et al. 2003). Recent studies have shown that 10–43% of all familial FTD patients are associated with mutations in the gene encoding the tau protein (*MAPT*) located at 17q21 (Hutton et al. 1998; Rizzu et al. 1999; Houlden et al. 1999; Poorkaj et al. 2001). To date, 36 different *MAPT* mutations have been identified in 106 dementia families worldwide (Rademakers et al. 2004a). However, in at least four autosomal-dominant FTD families, mutations in *MAPT* could not be identified despite conclusive linkage to 17q21 (Lendon et al. 1998; Rosso et al. 2001; Rademakers et al. 2002; Pickering-Brown et al. 2004). The recognition of an FTD locus at chromosome 17q21 independent of *MAPT* has stimulated many researchers to perform linkage analysis at 17q21 in extended FTD families with unknown genetic cause. This research has led to the identification of a group of at least nine families showing suggestive or conclusive linkage to 17q21 in the absence of detectable *MAPT* mutations (Bird et al. 1997; Froelich et al. 1997; Lendon et al. 1998; Kertesz et al. 2000; Rosso et al. 2001; Rademakers et al. 2002; Curcio et al. 2002; Pickering-Brown et al. 2004; Benussi et al. 2004). Clinically, there are no features that can reliably distinguish between patients from these nine families and *MAPT* mutation carriers, but, in contrast to FTD families segregating a *MAPT* mutation, these families are characterized by the absence of tau pathology. In fact, now that the number of FTD families linked to chromosome 17q21 is rapidly increasing and more neuropathological data become available, it seems as if the majority, if not all, of these FTD families belong to the FTD-U neuropathological subtype.

Here we present our most recent clinical, pathological and genetic data on the Dutch FTD-U family 1083 and three novel Belgian FTD families. In addition, our phenotypic and genetic findings will be discussed in relation to the fast increasing subgroup of 17q21-linked FTD families.

Family 1083

Family ascertainment and clinical description

Family 1083 is a four-generation, autosomal-dominant dementia family derived from a genetic epidemiological study of early-onset AD conducted in the Netherlands between 1980 and 1987 (Fig. 1; Hofman et al. 1989). This study included all patients diagnosed with clinical AD with an onset age at or before the age of 65 years, living in Rotterdam and the four northern provinces of the Netherlands. In 2001, an extensive clinical follow-up of family 1083 was conducted by neurological examination of patients and interviews of first-degree relatives. Available medical records and CT scans were also reviewed. At that time, 16 family members from three generations were considered affected (Rademakers et al. 2002). Recently, three additional family members developed dementia (III-10, IV-29 and IV-30). Detailed clinical information was available for 12 of the 19 patients. The mean age at onset in the 12 patients was 64.9 years (range 53 to 79 years), with mean disease duration of 6.6 years (range 3–13 years). In this family, the most common presenting symptoms were restlessness or roaming (eight patients), hyperorality (seven patients), and disinhibition (five patients). Language dysfunction consisted of echolalia or reduced spontaneous speech, eventually resulting in mutism in all patients. Other symptoms were memory problems early in the course of the disease in four patients and paranoia in three patients. Lobar atrophy on CT scan was present in two patients (frontal and temporal in III-29, and frontal in III-21), whereas the CT scan was normal one year after onset in III-11. Psychometric testing performed in patients III-11 and III-21 showed increased distractibility and impaired recall but intact recognition and spatial and temporal orientation. In conclusion, the clinical presentation was highly suggestive for FTD in eight patients, whereas AD and/or vascular dementia could not be excluded in four patients (III-22, III-23, III-26 and III-31).

Neuropathological description

Brain autopsy was performed in patient III-4. Macroscopic examination showed atrophy of frontal lobes with severely dilated ventricles (brain weight, 1,450 gram). Variable neuronal loss and microvacuolar changes were present in the superficial layers of the frontal cortex, with moderate gliosis of sub-pial and sub-cortical regions. In the temporal cortex, similar changes were seen in patches intermingled with normal cortex. Furthermore, mild neuronal loss was seen in the Cornu Ammonis of the hippocampus and the entorhinal cortex. A moderate number of age-related neurofibrillary tangles was present in the pyramidal cells of the entorhinal cortex, but no neurofibrillary degeneration was observed in any other cortical region. The parieto-occipital cortex and cerebellum were normal. Mild loss of pigmented neurons was seen in the substantia nigra, whereas striatal regions and thalamus were not affected. No neuritic or diffuse amyloid plaques, ballooned cells, Pick bodies or Lewy bodies were detected.

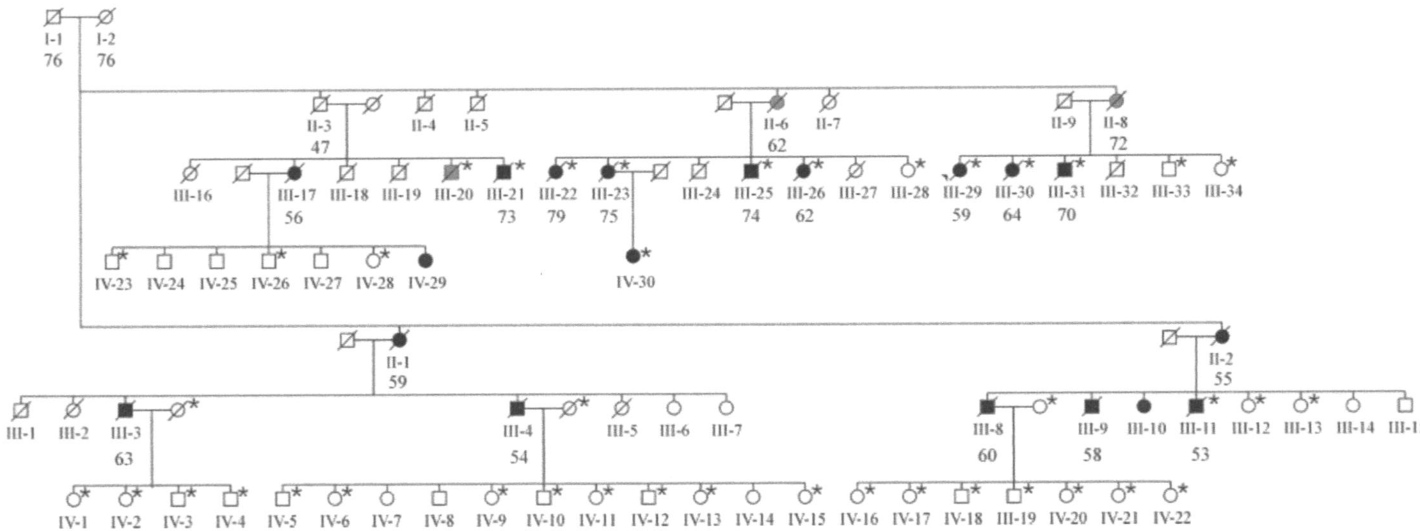

Fig. 1. Pedigree of family 1083. The numbering of the individuals is in accordance with van Duijn et al. (1994). Open symbols represent unaffected individuals or at-risk individuals with unknown phenotype, black symbols represent patients (n=19) with a diagnosis of FTD, AD and/or vascular dementia, and gray symbols represent patients (n=3) who suffered a stroke without reported signs of dementia. The numbers below the individuals denote age at onset for patients and age at death for obligate carriers. An arrowhead indicates the index patient. An asterisk (*) denotes that DNA was available for genotyping. Patient III-4 had pathological confirmation of FTD-U.

Immunohistochemical analysis

Staining with ubiquitin antibody (PC) showed small numbers of cytoplasmic neuronal inclusions and a moderate number of ubiquitin-positive neurites in the second layer of the frontal and temporal cortex. The inclusions were located predominantly in the perikaryal space directly surrounding the nucleus, and a few ubiquitin-positive inclusions appeared to be located within the nucleus and had a cat-eye or target shape. Neurons in the dentate gyrus and subcortical white matter did not contain any cytoplasmic ubiquitin-positive inclusions. Antibodies against tau (AT8, AT180, AT270, PHF1, MC1, BR01 and Tau2), α-synuclein (PC), neurofilament (SMI-32), β-tubulin (TUB 2.1), microtubule associated protein (MAP2), polyglutamine (1C2) and the prion protein ($PrP_{27\text{-}30}$) did not stain the inclusions. No amyloid plaques or ballooned cells were detected with βA4 antibody or αB-crystallin, respectively. Staining with GFAP antibody showed reactive astrocytosis in superficial cortical layers, as well as in subcortical white matter.

Genome-wide linkage analysis

Linkage to known AD loci on chromosomes 14, 19 and 21 was examined and linkage to 14q24.3, containing PSEN1 was excluded (van Duijn et al. 1994). In the index-patient (III-29), mutations in *APP* (exon 16 and 17), *PSEN1*, *PSEN2*, *PRNP* and *MAPT* (exons 9 to 13) were excluded by direct sequencing (Cruts et al. 1998; Roks et al. 1999; Dermaut et al. 2003). After the clinical follow-up study in 2001, DNA samples from 40 members of family 1083 were available for genetic analysis. Using the simulation program SLINK, we calculated that this family was informative enough to detect conclusive evidence for linkage with a simulated maximum LOD score of 4.63 and an average maximum LOD score of 1.80 at a recombination distance (θ) of 0.05 (Ott 1989; Weeks et al. 1990; Rademakers et al. 2002).

We conducted a 10-cM genome-wide scan with 400 microsatellite markers from the ABI Prism Linkage Mapping Set MD-10 Version 2 (Applied Biosystems, Foster City, CA). Two-point and multi-point LOD scores were calculated using MLINK and LINKMAP (LINKAGE 5.1; Lathrop et al. 1985). Only one of the 400 markers, D17S1868 at 17q21, showed a conclusive LOD score of 3.48 at $\theta = 0.00$. Linkage analysis with 22 additional chromosome 17 markers spanning a 32 cM region between D17S1857 and D17S787 on the Marshfield gender-average linkage map showed the highest two-point LOD score of 4.69 with D17S951 in the absence of recombinants. The two-point LOD-score at D17S951 further increased to 4.99 when patient IV-30, diagnosed with dementia after the original report in 2002, was included in the genetic analysis. A maximum multi-point LOD score of 5.56 was generated in the interval D17S1789-D17S951-D17S1860 on top of D17S951 ($\theta = 0$). Genotype data of 26 markers from the interval D17S1787 to D17S787 were used to reconstruct haplotypes taking into account minimal intermarker recombination. Genotype data in the offspring were used to reconstruct the haplotypes of deceased relatives. Segregation analysis showed that the 32-cM risk haplotype was present in all dementia patients, including the four patients for whom AD and/or vascular dementia could not be excluded. Two stroke patients (II-6 and II-

8) and individual II-3, who were obligate carriers, also segregated the risk haplotype. Individual II-3 died before the mean onset age in the family of an unrelated disease. The founders (I-1 and I-2) both died at age 76, with no reported history of dementia. Although most patients shared a common haplotype for all markers spanning the 32-cM region, four obligate recombinants were identified, defining a minimal candidate region of 4.8 cM between D17S1787 and D17S958.

Genomic *MAPT* mutation analysis

Because *MAPT* was located within the candidate region, and mutations in *MAPT* were responsible for the dementia in most of the families with chromosome 17-linked FTD with parkinsonism (FTDP-17), we performed an extensive mutation analysis of exons 1 to 14 of *MAPT* in seven patients and one control individual of family 1083 (Rademakers et al. 2002). In addition, we analyzed exon 0, more than 1 kb of 5' regulatory sequence and intron 13 that is retained in human *MAPT* transcripts. No mutations explaining the disease phenotype were observed. Segregation analysis of the different *MAPT* polymorphisms indicated that the risk-haplotype in family 1083 contained the common *MAPT* haplotype H1 defined by polymorphic alleles in exons 1, 2, 3, 9, 11 and 13. All patients were homozygous H1H1 except patients III-22 and III-26, who were heterozygous H1H2.

To exclude intronic and regulatory *MAPT* mutations, we initiated a sequencing project of the complete *MAPT* genomic region spanning over 135 kb (Rademakers et al. 2004b). Based on the genomic sequence with GenBank accession number NC_000017.9 from position 41323600 to 41462800, 192 primer sets were designed with SNPbox using the "saturation" module (Weckx et al. 2004a,b). These primer sets were amplified and sequenced in both directions in 23 individuals, including three patients and two control individuals from family 1083. During optimalization, 10 additional sequencing fragments were generated using Primer3 (Rozen and Skaletsky 2000). All sequence traces were analyzed by an in-house developed software tool, NovoSNP (Weckx et al., submitted for publication) and by visual inspection. In total we detected 574 single nucleotide polymorphisms and small insertions and deletions. Only one variant, g.38276T>A (numbering relative to AC091628.2), located in intron 0 of *MAPT* segregated with the disease haplotype in family 1083. Using a pyrosequencing assay, we analyzed 189 unrelated Dutch control individuals for this mutation and we identified 10 control individuals who were homozygous and 57 who were heterozygous for the A-allele, resulting in a minor allele frequency of 20.4%.

Belgian FTD families

Family ascertainment and clinical description

Index patients of families DR2, DR7 and DR8 were diagnosed with dementia and referred to the Molecular Diagnostic Unit in our department for routine DNA diagnostic screening. Mutation analysis of *APP* (exon 16 and 17), *PSEN1*, *PSEN2*,

Table 1. Clinical characteristics of three Belgian FTD families

Family	Number of generations	Number of patients	Mean onset age (range)	Mean disease duration (range)	DNA samples (patients)
DR2	5	12	65.7 (58-75)	5.4 (1-20)	38 (4)
DR8	4	11	60.3 (51-68)	7.4 (2-13)	37 (5)
DR7	4	8	71.0 (63-91)	5.2 (2-8)	15 (2)

PRNP and *MAPT* was performed and mutations were excluded. For each patient, detailed data on family history of dementia were collected, and additional patients and unaffected family members were asked to participate in a genetic study. Blood samples were obtained from 87 additional patients and relatives from DR2, DR7 and DR8, and DNA was extracted. The extended pedigrees are shown in Figure 2, and detailed clinical characteristics of the patients of each family are summarized in Table 1.

In family DR2, the most common presenting symptoms were severe language disturbances (II-1, III-3, III-4 and III-6) and changes of character (II-1, III-3, III-4, III-6 and III-8). In family DR8, detailed clinical information was available for two patients (III-18 and III-28), both showing typical frontal signs including personality changes, disinhibition, apathy, lack of insight, language problems, progressive memory problems and frontotemporal hypoperfusion on SPECT scan. In family DR7, similar to family DR8, aphasia was an important clinical finding (II-3, III-2, III-3 and III-4); in addition, at least three patients suffered from parkinsonian features (II-3, II-4 and III-3). In conclusion, the clinical presentation of 10 dementia patients from these three families was highly suggestive for FTD.

Neuropathological description

Brain autopsy was performed for patient III-4 of family DR7 (Fig. 3). No autopsy was available for patients of family DR2 and DR8. The brain of DR7/III-4 was grossly atrophied and weighed only ≈850 g, with severely dilated ventricles. This contrasted with patient III-4 of family 1083 described above, where the atrophy was not as severe. The atrophy mostly involved the frontal and temporal lobes and did not resemble that of Pick's disease, as the superior temporal gyrus was severely affected as well. The parieto-occipital cortex and cerebellum were normal, as described for the patient 1083/III-4. On microscopy of the left hemisphere, a severe microspongiosis and neuronal loss was most evident in the superior temporal gyrus. Despite this severe atrophy, we did not come across any neurofibrillary tangles or gross dystrophic neurites in any neocortical or hippocampal region analyzed. Senile plaques were also absent, although some degree of cerebral amyloid angiopathy (CAA) was present in the occipital leptomeninges. No ballooned cells, Pick bodies, or Lewy bodies were detected. Rarely, Marinesco bodies were observed in the substantia nigra.

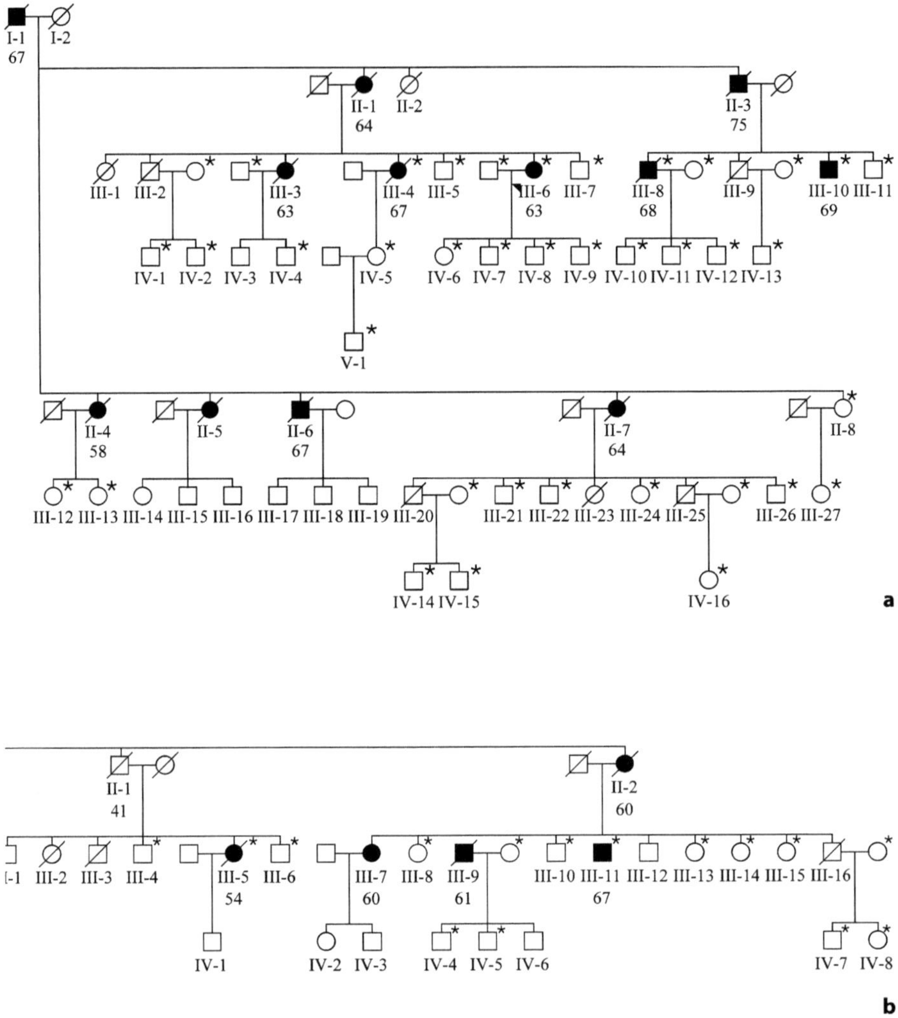

Fig. 2 a–c.

Immunohistochemical analysis

Most notably, ubiquitin-positive and mostly intranuclear inclusions were observed (Fig. 3), as described for the patient 1083/III-4. These inclusions were present in both neurons and glial cells in the cortex, neostriatum, and cerebellum. The morphology of these inclusions was typically globose, but on many occasions, they took the form of a cat's eye, as described for patient 1083/III-4. Notably, the neuronal inclusions were not positive for antibodies against tau (AT8, AT120, AT180, AT270; Innogenetics), α-synuclein (Chemicon), neurofilament (SMI-310),

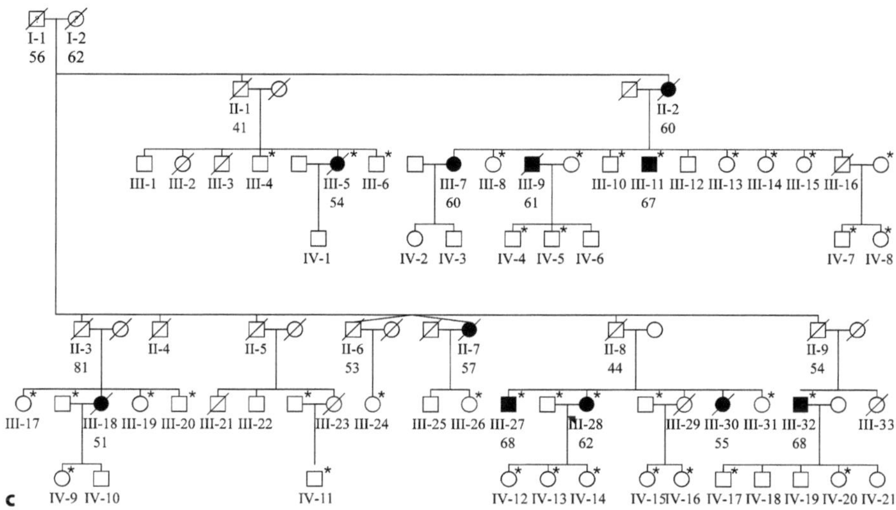

c

Fig. 2. Pedigrees of Belgian FTD families DR2 (**a**), DR7 (**b**) and DR8 (**c**). Open symbols represent unaffected individuals or at-risk individuals with unknown phenotype, black symbols represent dementia patients. The numbers below the individuals denote age at onset for patients and age at death for obligate carriers. An arrowhead indicates the index patient. An asterisk (*) denotes that DNA was available for genotyping. In family DR7, patient III-4 had pathological confirmation of FTD-U.

phosphorylated neurofilament (SMI-31), microtubule associated protein (MAP2), expanded polyglutamine (1C2), Huntington protein (HDJ-2), prion protein (3F4, Senetek), and GFAP (PAb, Dako). In addition, we are also seeking information on any involvement of specific elongation factors, heat shock proteins, aggregate-interacting proteins, or transcriptional factors. To this end, we have analyzed αB-crystalline (a chaperone of the small heat shock family; Novocasta), heat shock protein 40 (HDJ-2/DNAJ; Lab Vision), and TATA-bound TFIID (TFIID/TBP; Santa Cruz) without any significant staining. Many of these antibodies recognized expected pathology otherwise, for instance, αB-crystalline showed immunoreactivity in small oligodendroglial or microglial cells; AT270 stained axons in white matter; and TBP, very occasionally, stained astrocytic or microglial cytoplasm. No Aβ staining plaques were present although, as described by the classical stains, 4G8 detected some CAA in the leptomeninges. GFAP and microglial (HLA-DP, DQ, DR) reactivity was present in cortical and subcortical regions.

Chromosome 17-linkage analysis

DNA samples from 38 members of family DR2, 39 of family DR8 and 15 of family DR7 were available for genetic analysis. Simulation studies showed that families DR2 and DR8 were sufficiently informative to obtain conclusive linkage with simulated maximum LOD scores of 3.09 and 4.12 at a recombination distance (θ) of 0.00 for DR2 and DR8, respectively.

We first investigated whether families DR2 and DR8 were linked to the same chromosomal region at 17q21 that was identified in the Dutch FTD family 1083. We selected 18 markers spanning a region of 17.7 cM between D17S1863 and D17S1795, 15 markers from the Marshfield gender-averaged linkage map and three novel markers, chr17-16, chr17-19 and chr17-43, identified with the Tandem repeat finder Program (Table 2; Benson 1999). Chr17-16 is located in AC008105.32 starting at nt 82045, chr17-19 is located in AC091628.2 at nt 35879 and chr17-43 in AC068234.8 at nt 138430.

Two-point and multi-point LOD scores were calculated using MLINK and LINKMAP (LINKAGE 5.1; Lathrop et al. 1985). Family DR8 showed conclusive linkage to 17q21 with a maximum LOD score of 3.32 at marker D17S931 (θ = 0), whereas family DR2 only showed suggestive linkage with a maximum LOD score of 1.47 at markers chr17-43 (Table 2). Using multipoint linkage analysis,

Table 2. Two-point LOD scores at 17q21 in families DR2 and DR8.

Marker	Physical distance (Mb)	Genetic distance (cM)	LOD score at θ=0 DR2	DR8
D17S1863	0.0	0.0	−1.47	−0.02
D17S1818	8.4	9.7	0.20	2.26
D17S1814	9.7	10.7	0.81	0.00
D17S800	10.6	11.3	0.17	2.79
D17S1787	11.3	11.3	0.78	0.40
D17S1793	11.9	12.4	−0.06	0.67
D17S902	13.1	13.4	1.42	2.37
D17S951	13.4	12.9	0.30	2.79
D17S1861	14.4	12.9	1.01	2.47
D17S934	14.7	12.9	0.71	2.88
Chr17-16	15.0	–	−0.12	1.81
D17S810	15.2	12.9	0.38	0.59
Chr17-19	15.7	-	0.58	0.95
D17S920	16.3	13.4	0.70	0.41
D17S931	16.4	16.1	1.45	3.32
Chr17-43	16.8	–	1.47	1.76
D17S1785	18.1	16.1	0.47	2.60
D17S1795	19.4	17.7	−1.06	2.72

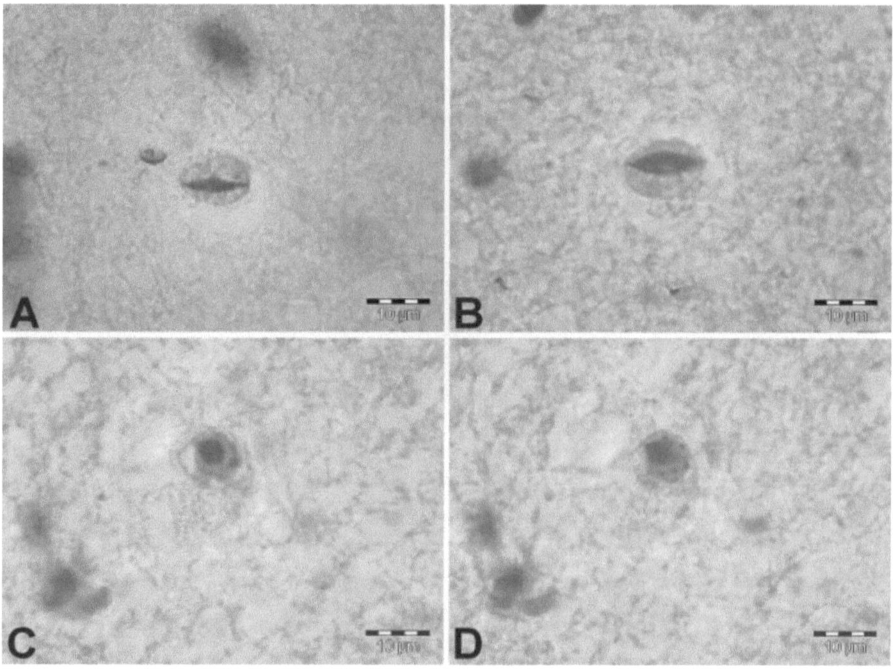

Fig. 3. Intranuclear ubiquitin-positive inclusion morphology in patient III-4 of family DR7. Intranuclear inclusions with typical cat-eye appearance were frequently observed in the superior temporal gyrus (**A**) as well as in the superior frontal gyrus (**B**). However, a globose appearance or other inclusion body morphologies were more commonly observed and were sometimes also accompanied by either a diffuse nuclear ubiquitin staining and/or additional smaller aggregates. Shown here are two planes of such an inclusion from the superior frontal gyrus region with one large and four to five small aggregates as well as a diffuse ubiquitin staining (**C** and **D**).

LOD scores increased to 3.49 in family DR8 and 1.79 in family DR2 in the interval D17S920-D17S931-D17S1795 on top of D17S931 ($\theta = 0$).

To confirm linkage to chromosome 17q21 in families DR2 and DR8 and to identify informative recombinants defining minimal candidate regions, haplotypes were reconstructed from genotype data of the 18 markers from the interval D17S1863 to D17S1795 (Fig. 4A and B). In family DR8, all patients carried the complete risk-haplotype, including individuals (II-1, II-3, II-8 and II-9) who were obligate carriers (Fig. 4B). Both founders (I-1 and I-2) died before the mean age at onset of dementia in the family. In family DR2, a risk-haplotype at 17q21 was also observed in all patients and obligate recombinants in two patients defined a candidate region of 17.7 cM between D17S1863 (III-4) and D17S1795 (III-6; Fig. 4A).

Moreover, detailed comparison of the linked alleles of STR markers at chromosome 17q21 showed that, in families DR2 and DR8, identical risk-haplotypes were segregating for all markers within the region flanked by D17S1814 and D17S1795 (Fig. 4A and B). These findings suggest that families DR2 and DR8 are genealogia

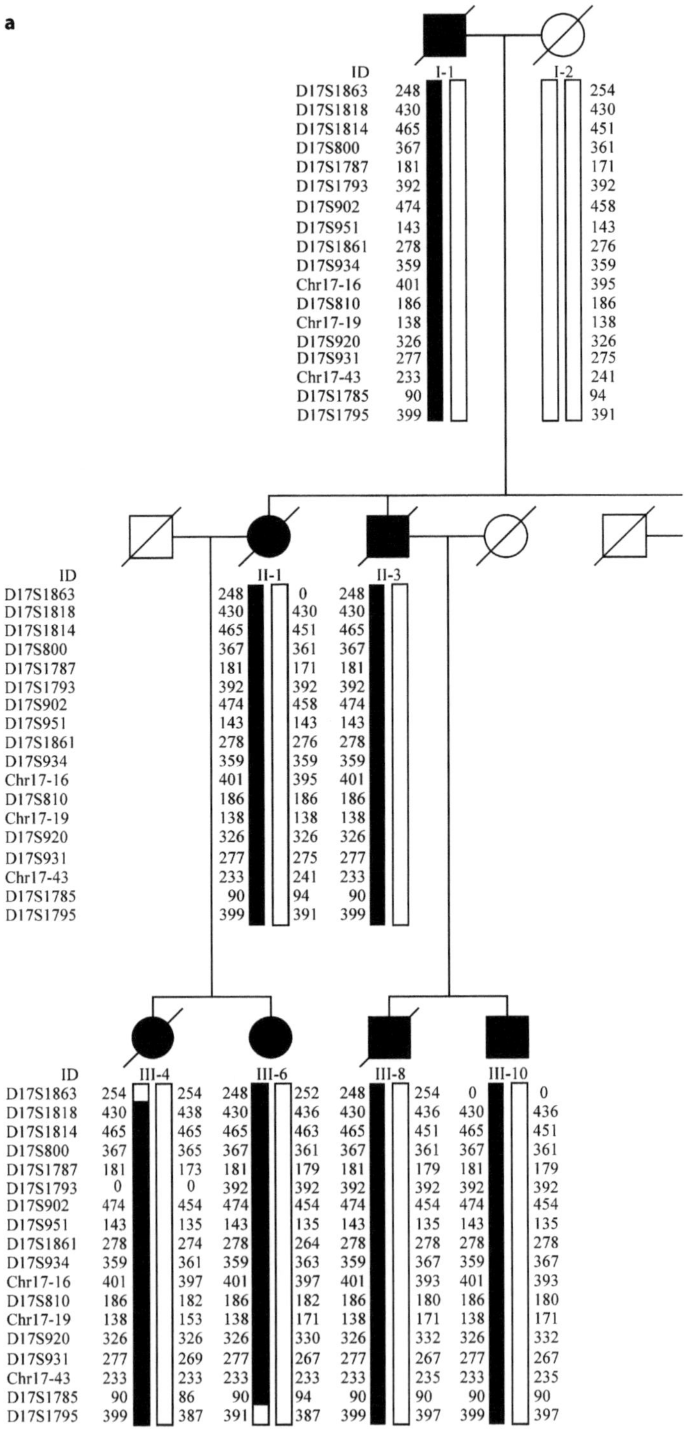
ID
I-1
I-2
D17S1863 248 254
D17S1818 430 430
D17S1814 465 451
D17S800 367 361
D17S1787 181 171
D17S1793 392 392
D17S902 474 458
D17S951 143 143
D17S1861 278 276
D17S934 359 359
Chr17-16 401 395
D17S810 186 186
Chr17-19 138 138
D17S920 326 326
D17S931 277 275
Chr17-43 233 241
D17S1785 90 94
D17S1795 399 391
ID
II-1
II-3
D17S1863 248 0 248
D17S1818 430 430 430
D17S1814 465 451 465
D17S800 367 361 367
D17S1787 181 171 181
D17S1793 392 392 392
D17S902 474 458 474
D17S951 143 143 143
D17S1861 278 276 278
D17S934 359 359 359
Chr17-16 401 395 401
D17S810 186 186 186
Chr17-19 138 138 138
D17S920 326 326 326
D17S931 277 275 277
Chr17-43 233 241 233
D17S1785 90 94 90
D17S1795 399 391 399
ID
III-4
III-6
III-8
III-10
D17S1863 254 254 248 252 248 254 0 0
D17S1818 430 438 430 436 430 436 430 436
D17S1814 465 465 465 463 465 451 465 451
D17S800 367 365 367 361 367 361 367 361
D17S1787 181 173 181 179 181 179 181 179
D17S1793 0 0 392 392 392 392 392 392
D17S902 474 454 474 454 474 454 474 454
D17S951 143 135 143 135 143 135 143 135
D17S1861 278 274 278 264 278 278 278 278
D17S934 359 361 359 363 359 367 359 367
Chr17-16 401 397 401 397 401 393 401 393
D17S810 186 182 186 182 186 180 186 180
Chr17-19 138 153 138 171 138 171 138 171
D17S920 326 326 326 330 326 332 326 332
D17S931 277 269 277 267 277 267 277 267
Chr17-43 233 233 233 233 233 235 233 235
D17S1785 90 86 90 94 90 90 90 90
D17S1795 399 387 391 387 399 397 399 397

b

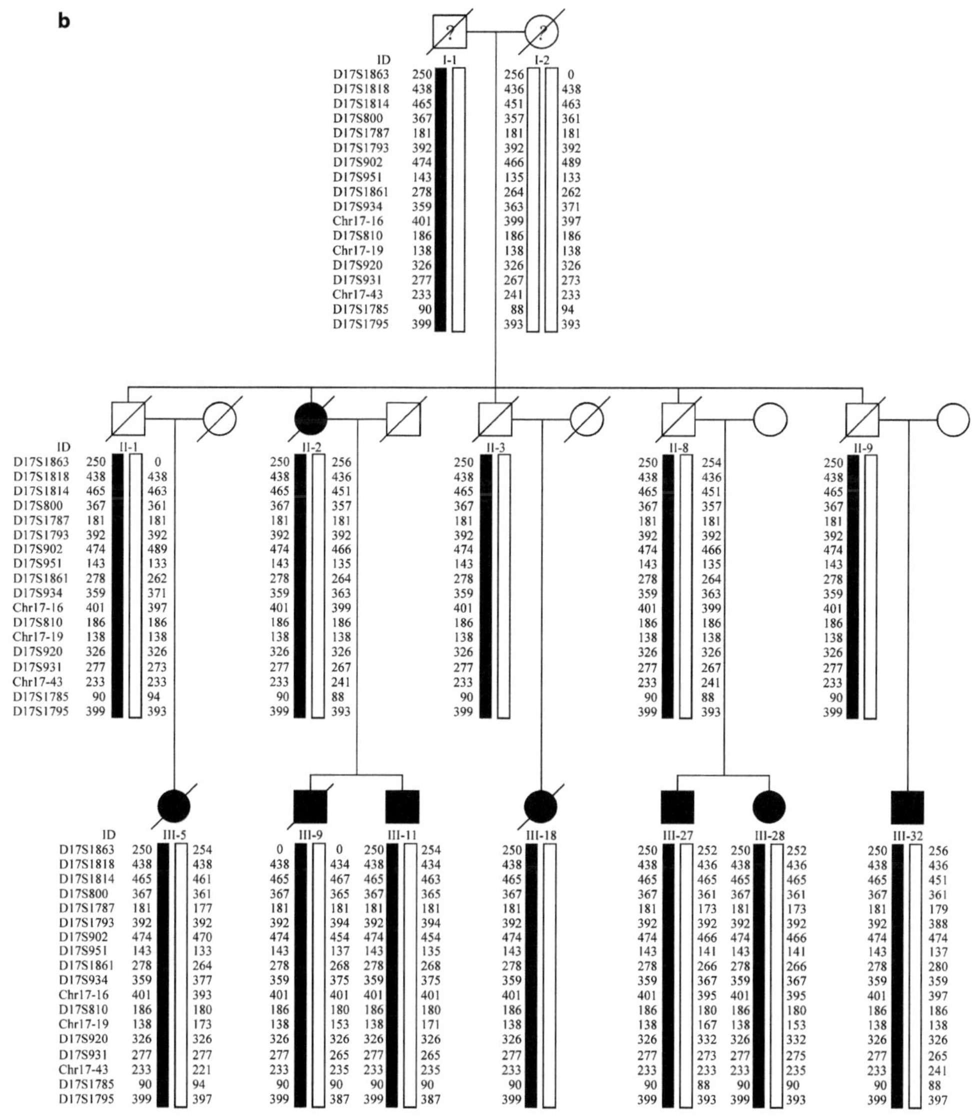

Fig. 4. Linkage pedigrees of families DR2 (A) and DR8 (B) showing haplotypes based on 18 informative markers at 17q21. Only patients and obligate carriers used in the linkage analysis are shown. Haplotypes for individuals from the first and second generation as well as for patients III-9 and III-18 from family DR8 were reconstructed through offspring. In family DR8, the risk haplotype was arbitrarily set for I-1.

cally related and originated from a common founder. Therefore, by combining the segregation data of families DR2 and DR8, we were able to reduce the minimal candidate region in these families to 8.04 cM between D17S1818 and D17S1795, delineated by haplotype sharing at the centromeric site and a recombinant in family DR2 (III-4) at the telomeric site (Fig. 4A).

No obvious haplotype sharing was observed between the risk-haplotypes of families DR2 and DR8 and Dutch family 1083. Also, *MAPT* mutation analysis in additional family members of families DR2 and DR8 showed that patient III-32 from family DR8 was homozygous for the extended *MAPT* haplotype H2, implying that the mutation responsible for FTD in DR8 is located on *MAP*T H2 whereas the mutation in family 1083 is located on *MAP*T H1.

Conclusions

In this study, we showed the results of a genome-wide search in a four-generation Dutch pedigree with autosomal-dominant early-onset dementia. Clinical and neuropathological follow-ups of the family showed that the phenotype most closely resembled FTD-U. In this family, we provided significant linkage evidence for a gene at 17q21 and reduced the candidate region to 4.8 cM between D17S1787 and D17S958 comprising *MAPT*. Using a genomic sequencing approach, mutations in the complete *MAPT* genomic region, including 5 kb of regulatory sequence located upstream of exon 0, were excluded.

In addition, we presented three novel Belgian FTD families showing clinical and pathological characteristics similar to family 1083. Autopsy was only available for family DR7, showing ubiquitin-positive, tau-negative cytoplasmic and predominantly nuclear inclusions leading to the pathological sub classification of FTD-U. Due to the limited number of DNA samples available for this family, linkage analysis was not performed. In contrast, families DR8 and DR2 were informative and showed LOD scores of 3.49 and 1.79 at D17S931, respectively. Since identical alleles were present on the 17q21 risk-haplotypes segregating in these families, it was suggested that families DR2 and DR8 are genealogically related, originating from a common founder. The candidate region identified in these families was 8.04 cM between D17S1818 and D17S1795 and completely overlapped with that previously identified in family 1083. The identification of two novel chromosome 17-linked FTD families in Belgium emphasizes the importance of this FTD locus in the general population, and identification of the underlying gene defect is expected to provide new insights into the characteristics of this FTD subtype.

Comparison of chromosome 17-linked, tau-negative FTD families

To date, 11 FTD families have been linked to chromosome 17q21 without detectable *MAPT* mutations. The genetic, clinical and pathological data from these 11 families are summarized in Table 3. In five of these families, conclusive linkage to 17q21 was detected (1083, DR8, HDDD2, UBC-17 and Dutch III; Lendon et al. 1998; Rosso et al. 2001; Rademakers et al. 2002; Pickering-Brown et al. 2004). Family 1083, described in detail in this study, is the most informative of these families, with a multi-point LOD score of 5.56 at D17S951, and reduces the candidate region to a 4.8 cM interval between D17S1787 and D17S958 (Fig. 5). Centromeric and/or telomeric boundaries of the candidate region were also identified by informative recombinants in the other conclusively 17-linked families, defining a minimal region between D17S1787 and D17S931 shared by all families (Fig. 5). Although both markers showing telomeric recombinants D17S958 (1083) and D17S931 (UBC-17) are mapped at identical positions on genetic maps, there is a physical distance of 1.2 Mb between these markers, substantially reducing the candidate region to approximately 5.4 Mb. Unfortunately, according to the NCBI genome build 34.3, at least 136 known genes are located within this candidate interval, including *MAPT*.

Thus far, mutations in *MAPT* have only been identified in FTDP-17 families with pathological tau depositions (for review, see Rademakers et al. 2004a). All FTD families described in Table 3 had no distinctive tau pathology, in agreement with the absence of mutations in *MAPT*. Although a more complex mutation in *MAPT*, unrecognized by standard sequencing procedures, cannot be entirely excluded, the absence of tau pathology is suggestive of another defective gene at 17q21 responsible for FTD. Also, the identification of 17-linked families segregating the extended *MAPT* haplotype H1 as part of their risk-haplotype (1083) while others are segregating H2 (DR2 and DR8) indicates that the presence of one of

Table 3. Genetic, clinical and pathological characteristics of tau-negative, chromosome 17-linked FTD families

Family	Maximum LOD score	Mean onset age in years (range)	Mean disease duration in years (range)	FTD-U pathology	Reference
1083	5.56	65 (53-79)	7 (3-13)	+	Rademakers et al. 2002 and this study
HDDD2	3.68	60 (52-77)	7 (3-11)		Lendon et al. 1998
UBC-17	3.64	59 (42-72)	7 (3-14)	+	Pickering-Brown et al. 2004
DR8	3.49	60 (51-68)	7 (2-13)		This study
Dutch-III	3.46	61 (53-71)	9 (6-13)	+	Rosso et al. 2001
Brescia-F071	2.90	62 (52-75)	ND		Benussi et al. 2004
Karolinska	2.86	51 (48-55)	3 (2-4)	+	Froelich et al., 1997
Calabrian	2.70	63 (35-78)	4 (1-12)		Curcio et al. 2002
DR2	1.79	66 (58-75)	6 (1-20)		This study
Kertesz	1.68	43 (38-47)	8 (5-10)	+	Kertesz et al. 2000
Seattle B	1.11	55 (ND)	10 (ND)		Bird et al. 1997

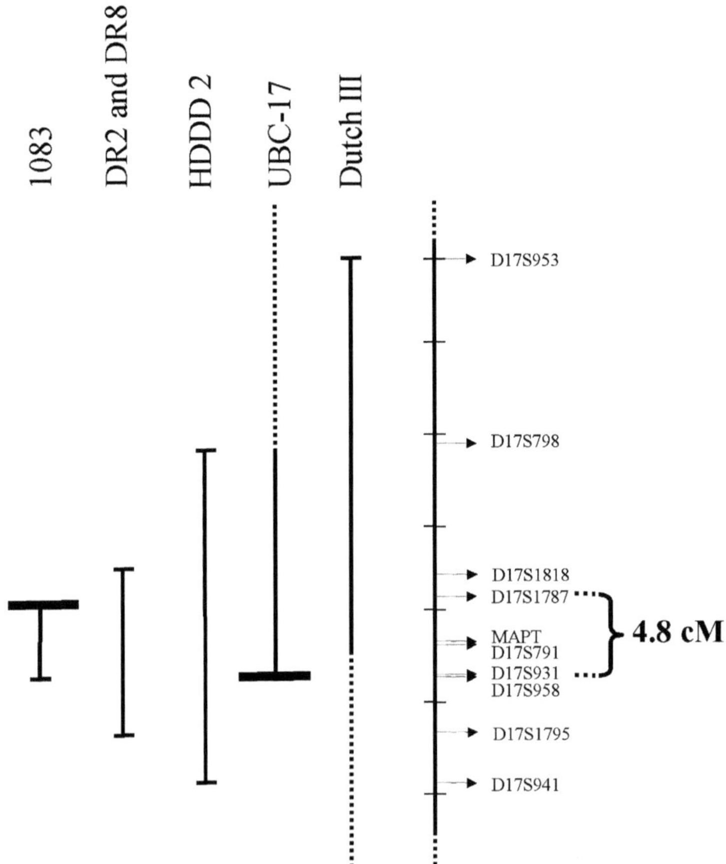

Fig. 5. Schematic presentation of the candidate region in conclusively 17q21-linked FTD families without *MAPT* mutations. Dashed lines indicate that the disease gene could be in a region that extends beyond the map shown, or that genetic linkage analysis with markers from this chromosomal region was not performed.

these haplotypes per se is not contributing to the disease mechanism. The mean onset age, disease duration and mean age at death in the Dutch family 1083 and the Belgian families DR2 and DR8 that were reported in this study were most comparable with families HDDD2, Dutch III, and UBC-17 (Table 4; Lendon et al. 1998; Rosso et al. 2001; Pickering-Brown et al. 2004). In the Karolinska, Kertesz and Seattle B families, onset ages were 10 to 20 years earlier, with narrower onset age ranges particularly in the Karolinska and Kertesz families (Bird et al. 1997; Froelich et al. 1997; Kertesz et al. 2000). Also, an extremely rapid progression of disease was reported for the Karolinska family and the Calabrian kindred, with an average disease duration of only three to four years (Bird et al. 1997; Froelich et al. 1997; Basun et al. 1997; Curcio et al. 2002). The identification of similar clinical characteristics, especially within the group of conclusively 17-linked FTD families, strengthens the hypothesis that mutations in a single gene might be responsible for the dementia in these different FTD families.

In addition to the complete lack of tau neuropathology, the observation of ubiquitin-positive neuronal inclusions, characteristic of FTD-U, was the most important neuropathological finding in at least three of the conclusively 17-linked families (1083, Dutch III, UBC17; Rosso et al. 2001; Rademakers et al. 2002; Pickering-Brown et al. 2004). Ubiquitin-positive inclusions were also reported in the Kertesz family and in at least one patient from the Karolinska family (Kertesz et al. 2000; Rosso et al. 2001; Froelich et al. 2003). Assigning family HDDD2 to one of the three broad pathological FTD subtypes remains controversial, since many different studies have reported on the neuropathological findings in this family and detailed immunohistochemical ubiquitin staining was not described (Lendon et al. 1998; Zhukareva et al. 2001, 2003).

Differentiating FTD-U from DLDH has been the subject of several recent neuropathological studies (Hodges et al. 2004; Paviour et al. 2004; Josephs et al. 2004; Lipton et al. 2004). Although DLDH was long thought to be the most common pathology underlying the clinical diagnosis of FTD, at least two recent reports emphasized the importance of the FTD-U neuropathological subtype. In a study of 76 neuropathologically confirmed FTD patients, Lipton et al. (2004) diagnosed 38% of the patients as FTD-U, including three patients previously reported as DLDH. None of their 76 patients was ultimately diagnosed as DLDH. In a recent study by Josephs et al. (2004), the revised prevalence of FTD-U increased from 45 to 62%, whereas the prevalence of DLDH was only 3% compared to 20% estimated previously. The authors argued that the diagnosis of DLDH should be made with care and only after exhaustive immunohistochemical examination. Interestingly, in addition to cytoplasmic neuronal inclusions, "cat-eye" shaped intranuclear neuronal inclusions were identified in families 1083, Dutch III and UBC-17 (Rosso et al. 2001; Pickering-Brown et al. 2004; Rademakers et al. 2004a). These intranuclear inclusions might be a pathological hallmark characteristic of FTD patients from families carrying a specific genetic defect at 17q21. In this view, the finding that neuronal intranuclear inclusions distinguish familial FTD from sporadic FTD patients, as shown by Mackenzie and Feldman (2004), might be relevant.

In conclusion, the absence of tau pathology in the majority of FTD patients, together with the relatively low *MAPT* mutation frequency in familial FTD, suggests that other pathogenic mechanisms and defective genes resulting in FTD without tauopathy are to be expected. Genetic locus heterogeneity in FTD was already illustrated by a Danish FTD family linked to a 12 cM interval in the centromeric region of chromosome 3 (Brown et al. 1995) and by linkage to a 17 cM region at 9q21-22 reported in a set of families in which patients developed amyotrophic lateral sclerosis (ALS) and FTD, or either ALS or FTD alone (Hosler et al. 2000). However, no additional families have been conclusively linked to these loci, possibly reflecting the relatively low contribution of these genetic loci to familial FTD. In contrast, five FTD families have now been described with conclusive linkage to 17q21 in the absence of *MAPT* mutations and at least six families showed suggestive linkage to the same chromosomal region (Table 3). Since pathological brain examination showed ubiquitin-positive inclusions at brain autopsy in patients of at least five of these families, and since several studies recently reported that FTD-U might be the most important neuropathological FTD subtype, we speculate that the gene responsible for the dementia in these 17-linked families might make an important genetic contribution to the development of FTD in the general population. Furthermore, it is expected that the identification of this gene will improve our understanding of the genetic etiology of FTD as well as our understanding of the pathogenesis of FTD and its relation to other neurodegenerative disorders.

Acknowledgments

The authors are grateful to Karin Peeters for genealogical studies and sampling of the Belgian FTD-U families, to Tim De Pooter and Marleen Van den Broeck for support with the genetic studies, and to the personnel of the VIB8 Genetic Service Facility (http://www.vibgeneticservicefacility.be/) for the genomic sequencing of *MAPT*. The research described in this paper was supported by the Special Research Fund of the University of Antwerp (BOF-UA), the Fund for Scientific Research Flanders (FWO-F), the Interuniversity Attraction Poles (IUAP) program P5/19 of the Belgian Federal Science Policy Office, the Medical Foundation Queen Elisabeth (GSKE), the International Alzheimer Research Foundation (IARF), Belgium, and the EU contract LSHM-CT-2003-503330 (APOPIS) R.R. and M.C. are postdoctoral fellows of the FWO-F.

References

Basun H, Almkvist O, Axelman K, Brun A, Campbell TA, Collinge J, Forsell C, Froelich S, Wahlund LO, Wetterberg L, Lannfelt L (1997) Clinical characteristics of a chromosome 17-linked rapidly progressive familial frontotemporal dementia. Arch Neurol 54: 539–544.

Benson G (1999). Tandem repeats finder: a program to analyze DNA sequences. Nucleic Acids Res 27: 573–580.

Benussi L, Signorini S, Ghidoni R, Alberici A, Barbiero L, Feudatari E, Gigola L, Nicosia F, Paterlini A, Strom T, Bettecken T, Meitinger T, Binetti G (2004) Identification of genetic loci associated with familial frontotemporal dementia. Neurobiol Aging 25: P4–149.

Bird TD, Wijsman EM, Nochlin D, Leehey M, Sumi SM, Payami H, Poorkaj P, Nemens E, Rafkind M, and Schellenberg GD (1997) Chromosome 17 and hereditary dementia: linkage studies in three non- Alzheimer families and kindreds with late-onset FAD. Neurology 48: 949–954.

Brown J, Ashworth A, Gydesen S, Sorensen A, Rossor M, Hardy J, Collinge J (1995) Familial non-specific dementia maps to chromosome 3. Human Mol Genet 4: 1625–1628.

Cruts M, van Duijn CM, Backhovens H, Van den BM, Wehnert A, Serneels S, Sherrington R, Hutton M, Hardy J, George-Hyslop PH, Hofman A,Van Broeckhoven C (1998). Estimation of the genetic contribution of presenilin-1 and -2 mutations in a population-based study of presenile Alzheimer disease. Human Mol Genet 7: 43–51.

Curcio SA, Kawarai T, Paterson AD, Maletta RG, Puccio G, Perri M, Di Natale M, Palermo S, Foncin JF, Hyslop PH, Bruni AC (2002) A large Calabrian kindred segregating frontotemporal dementia. J Neurol 249: 911–922.

Dermaut B, Croes EA, Rademakers R, Van den BM, Cruts M, Hofman A, van Duijn CM, Van Broeckhoven C (2003) PRNP Val129 homozygosity increases risk for early-onset Alzheimer's disease. Ann Neurol 53: 409–412.

Froelich FS, Axelman P, Almkvist A, Basun H, Lannfelt L (2003) Extended investigation of tau and mutation screening of other candidate genes on chromosome 17q21 in a Swedish FTDP-17 family. Am J Med Genet 121B: 112–118.

Froelich S, Basun H, Forsell C, Lilius L, Axelman K, Andreadis A, Lannfelt L (1997) Mapping of a disease locus for familial rapidly progressive frontotemporal dementia to chromosome 17q12-21. Am J Med Genet 74: 380–385.

Hodges JR, Davies R, Xuereb J, Kril J, Halliday G (2003) Survival in frontotemporal dementia. Neurology 61: 349–354.

Hodges JR, Davies RR, Xuereb JH, Casey B, Broe M, Bak TH, Kril JJ, Halliday GM (2004) Clinicopathological correlates in frontotemporal dementia. Ann Neurol 56: 399–406.

Hofman A, Schulte W, Tanja TA, van Duijn CM, Haaxma R, Lameris AJ, Otten VM, Saan RJ (1989) History of dementia and Parkinson's disease in 1st-degree relatives of patients with Alzheimer's disease. Neurology 39: 1589–1592.

Hosler BA, Siddique T, Sapp PC, Sailor W, Huang MC, Hossain A, Daube JR, Nance M, Fan C, Kaplan J, Hung WY, McKenna-Yasek D, Haines JL, Pericak-Vance MA, Horvitz HR, Brown RH, Jr. (2000) Linkage of familial amyotrophic lateral sclerosis with frontotemporal dementia to chromosome 9q21-q22. JAMA 284: 1664–1669.

Houlden H, Baker M, Adamson J, Grover A, Waring S, Dickson D, Lynch T, Boeve B, Petersen RC, Pickering-Brown S, Owen F, Neary D, Craufurd D, Snowden J, Mann D, Hutton M (1999) Frequency of tau mutations in three series of non-Alzheimer's degenerative dementia. Ann Neurol 46: 243–248.

Hutton M, Lendon CL, Rizzu P, Baker M, Froelich S, Houlden H, Pickering-Brown S, Chakraverty S, Isaacs A, Grover A, Hackett J, Adamson J, Lincoln S, Dickson D, Davies P, Petersen RC, Stevens M, de Graaff E, Wauters E, van Baren J, Hillebrand M, Joosse M, Kwon JM, Nowotny P, CheLK, Norton J, Morris JC, Reed LA, Trojanowski J, Basun H, Lannfelt L, Neystat M, Fahn S, Dark F, Tannenberg T, Dodd PR, Hayward N, Kwok JBJ, Schofield PR, Andreadis A, Snowden J, Craufurd D, Neary D, Owen F, Oostra BA, Hardy J, Goate A, van Swieten J, Mann D, Lynch T, Heutink P (1998) Association of missense and 5'-splice-site mutations in tau with the inherited dementia FTDP-17. Nature 393: 702–705.

Josephs KA, Holton JL, Rossor MN, Godbolt AK, Ozawa T, Strand K, Khan N, Al Sarraj S, Revesz T (2004) Frontotemporal lobar degeneration and ubiquitin immunohistochemistry. Neuropathol Appl Neurobiol 30: 369–373.

Kertesz A, Kawarai T, Rogaeva E, George-Hyslop P, Poorkaj P, Bird TD, Munoz DG (2000) Familial frontotemporal dementia with ubiquitin-positive, tau-negative inclusions. Neurology 54: 818–827.

Lathrop GM, Lalouel JM, Julier C, Ott J (1985) Multilocus linkage analysis in humans: detection of linkage and estimation of recombination. Am J Hum Genet 37: 482–498.

Lendon CL, Lynch T, Norton J, McKeel DW, Jr., Busfield F, Craddock N, Chakraverty S, Gopalakrishnan G, Shears SD, Grimmett W, Wilhelmsen KC, Hansen L, Morris JC, Goate AM (1998) Hereditary dysphasic disinhibition dementia: a frontotemporal dementia linked to 17q21-22. Neurology 50: 1546–1555.

Lipton AM, White CL, III, Bigio EH (2004) Frontotemporal lobar degeneration with motor neuron disease-type inclusions predominates in 76 cases of frontotemporal degeneration. Acta Neuropathol (Berl).

Mackenzie IR, Feldman H (2004) Neuronal intranuclear inclusions distinguish familial FTD–MND type from sporadic cases. Dement Geriatr Cogn Disord 17: 333–336.

McKhann GM, Albert MS, Grossman M, Miller B, Dickson D, Trojanowski JQ (2001) Clinical and pathological diagnosis of frontotemporal dementia: report of the Work Group on Frontotemporal Dementia and Pick's Disease. Arch Neurol 58: 1803–1809.

Morris HR, Khan MN, Janssen JC, Brown JM, Perez-Tur J, Baker M, Ozansoy M, Hardy J, Hutton M, Wood NW, Lees AJ, Revesz T, Lantos P, Rossor MN (2001) The genetic and pathological classification of familial frontotemporal dementia. Arch Neurol 58: 1813–1816.

Neary D, Snowden JS, Gustafson L, Passant U, Stuss D, Black S, Freedman M, Kertesz A, Robert PH, Albert M, Boone K, Miller BL, Cummings J, Benson DF (1998) Frontotemporal lobar degeneration: a consensus on clinical diagnostic criteria. Neurology 51: 1546–1554.
Ott J (1989) Computer-simulation methods in human linkage analysis. Proc Natl Acad Sci USA 86: 4175–4178.
Paviour DC, Lees AJ, Josephs KA, Ozawa T, Ganguly M, Strand C, Godbolt A, Howard RS, Revesz T, Holton JL (2004) Frontotemporal lobar degeneration with ubiquitin-only-immunoreactive neuronal changes: broadening the clinical picture to include progressive supranuclear palsy. Brain. 127: 2441–2451
Pickering-Brown S, Baker M, Adamson J, Melquist S, Nyborg A, Lindholm C, Hutton M,Mackenzie I (2004) Familial frontotemporal dementia with neuronal intranuclear inclusions is linked to chromosome 17. Soc Neurosci Abstr.97.9.
Poorkaj P, Grossman M, Steinbart E, Payami H, Sadovnick A, Nochlin D, Tabira T, Trojanowski JQ, Borson S, Galasko D, Reich S, Quinn B, Schellenberg G, Bird TD (2001) Frequency of tau gene mutations in familial and sporadic cases of non- Alzheimer dementia. Arch Neurol 58: 383–387.
Rademakers R, Cruts M, Dermaut B, Sleegers K, Rosso SM, Van den BM, Backhovens H, Van Swieten J, van Duijn CM, Van Broeckhoven C (2002) Tau negative frontal lobe dementia at 17q21: significant finemapping of the candidate region to a 4.8 cM interval. Mol Psychiatr 7: 1064–1074.
Rademakers R, Cruts M, Van Broeckhoven C (2004a) The role of tau (MAPT) in frontotemporal dementia and related tauopathies. Human Mutat 24: 277–295.
Rademakers R, van der Zee, J, Bogaerts V, Van den Bossche D, Backhovens H, De Pooter T, Bel Kacem S, van Duijn C, Del-Favero J, Van Broeckhoven C (2004b) Genomic sequencing of MAPT provides an extended SNP map and identifies >30 H1 subhaplotypes. Neurobiol Aging 25: P4–154.
Ratnavalli E, Brayne C, Dawson K, Hodges JR (2002) The prevalence of frontotemporal dementia. Neurology 58: 1615–1621.
Rizzu P, van Swieten JC, Joosse M, Hasegawa M, Stevens M, Tibben A, Niermeijer MF, Hillebrand M, Ravid R, Oostra BA, Goedert M, van Duijn CM, Heutink P (1999) High prevalence of mutations in the microtubule-associated protein tau in a population study of frontotemporal dementia in the Netherlands. Am J Human Genet 64: 414–421.
Roks G, Dermaut B, Heutink P, Julliams A, Backhovens H, Van de BM, Serneels S, Hofman A, Van Broeckhoven C, van Duijn CM, Cruts M (1999) Mutation screening of the tau gene in patients with early-onset Alzheimer's disease. Neurosci Lett 277: 137–139.
Rosso SM, Donker KL, Baks T, Joosse M, de K, I, Pijnenburg Y, de Jong D, Dooijes D, Kamphorst W, Ravid R, Niermeijer MF, Verheij F, Kremer HP, Scheltens P, van Duijn CM, Heutink P, van Swieten JC (2003) Frontotemporal dementia in The Netherlands: patient characteristics and prevalence estimates from a population-based study. Brain 126: 2016–2022.
Rosso SM, Kamphorst W, de Graaf B, Willemsen R, Ravid R, Niermeijer MF, Spillantini MG, Heutink P, van Swieten JC (2001) Familial frontotemporal dementia with ubiquitin-positive inclusions is linked to chromosome 17q21–22. Brain 124: 1948–1957.
Rozen S, Skaletsky H (2000) Primer3 on the WWW for general users and for biologist programmers. Meth Mol Biol 132: 365–386.
Stevens M, van Duijn CM, Kamphorst W, de Knijff P, Heutink P, van Gool WA, Scheltens P, Ravid R, Oostra BA, Niermeijer MF, van Swieten JC (1998) Familial aggregation in frontotemporal dementia. Neurology 50: 1541–1545.
van Duijn CM, Hendriks L, Farrer LA, Backhovens H, Cruts M, Wehnert A, Hofman A, Van Broeckhoven C (1994) A population-based study of familial Alzheimer disease: linkage to chromosomes 14, 19, and 21. Am J Human Genet 55: 714–727.
Weckx S, De Rijk P, Van Broeckhoven C, Del Favero J (2004a). SNPbox, a modular software package for large scale primer design. Bioinformatics doi: 10.1093

Weckx S, De Rijk P, Van Broeckhoven C, Del Favero J (2004b) SNPbox: web-based high-throughput primer design from gene to genome. Nucleic Acids Res 32: W170-W172.

Weeks, Ott J, Lathrop GM (1990) SLINK: a general simulation program for linkage analysis. Am J Human Genet 47: A204.

Zhukareva V, Vogelsberg-Ragaglia V, Van Deerlin VM, Bruce J, Shuck T, Grossman M, Clark CM, Arnold SE, Masliah E, Galasko D, Trojanowski JQ, Lee VM (2001) Loss of brain tau defines novel sporadic and familial tauopathies with frontotemporal dementia. Ann Neurol 49: 165–175.

Zhukareva V, Sundarraj S, Mann D, Sjogren M, Blenow K, Clark CM, McKeel DW, Goate A, Lippa CF, Vonsattel JP, Growdon JH, Trojanowski JQ, Lee VM (2003) Selective reduction of soluble tau proteins in sporadic and familial frontotemporal dementias: an international follow-up study. Acta Neuropathol (Berl) 105: 469–476.

Variations of the Phenotype in Frontotemporal Dementias

N. L. Foster [1*], *R. E. Koeppe* [2], *B. J. Giordani* [3], *T. E. Nichols* [4], *A. P. Lieberman* [5]

Summary

To provide better care for patients, physicians most often seek to determine the cause of symptoms by recognizing a clinical phenotype and then relating that phenotype to known disease mechanisms, including genetic variations and pathology. This approach has been particularly difficult with frontotemporal dementias (FTD), because of the diversity and continuous evolution of its behavioral, language and cognitive symptoms. Recent systematic clinical-pathological observations have resulted in proposed clinical criteria for FTD, but accurate diagnosis remains challenging and, in clinical practice, the phenotype of FTD can easily be confused with other disorders.

Positron emission tomography with [^{18}F]fluorodeoxyglucose (FDG-PET) can aid in the recognition of clinical phenotype and explore disease mechanisms. We reviewed the results of FDG-PET in 19 patients who subsequently had FTD confirmed at postmortem examination, including one with a known tau gene mutation. A comparison of scans from normal elderly controls with scans of patients with pathologically confirmed Alzheimer's disease (AD) demonstrates that FTD causes a distinctive pattern of hypometabolism, with considerable individual variability within this general pattern. FTD consistently causes predominant hypometabolism in frontal association, anterior cingulate, and anterior temporal regions, but the involvement of each region is variable and can be either symmetric or asymmetric. As the illness progresses, glucose hypometabolism becomes more pervasive, extending into regions characteristically affected early in AD. Although FDG-PET abnormalities accurately reflect clinical phenotype, neither pathological diagnosis nor genotype reliably predicts the variations observed in the pattern of glucose hypometabolism in FTD.

[1] Department of Neurology, [2] Department of Radiology, [3] Department of Psychiatry, [4] Department of Biostatistics, [5] Department of Pathology, University of Michigan, Ann Arbor, MI, USA, 48109

* Corresponding Author: Norman L. Foster, M.D., Department of Neurology, University of Michigan, 1920 Taubman Center, 1500 E. Medical Center Dr., Ann Arbor, MI 48109-0316, Phone: 734-936-9045, FAX: 734-936-8763, email: nlfoster@umich.edu

Cummings et al.
Genotype – Protéotype – Phenotype
Relationships in Neurodegenerative Diseases

Introduction

It is becoming increasingly important to recognize the specific cause of dementia as we learn more about neurodegenerative disorders. Dementing diseases have different risk factors, courses, prognoses, and pathophysiologies, and consequently, their appropriate treatment also differs. Physicians currently rely on a medical history and examination to identify a clinical phenotype and make a diagnosis. This approach is difficult, because it requires subjective interpretation of information from diverse sources and often depends upon the reliability and insight of family members. Furthermore, the clinical phenotype of progressive dementing diseases evolves over time, and distinctive features early in the illness may be obscured as functional abilities decline and patients are less able to cooperate. Positron emission tomography with [^{18}F]fluorodeoxyglucose (FDG-PET) offers a more objective approach to aid in the recognition of phenotype. Because of the current fallibility of clinical diagnosis, a better understanding of the metabolic changes seen with FDG-PET must be grounded in the study of autopsy-confirmed cases.

Frontotemporal dementia (FTD) is a particularly good example of the relationship between phenotype, proteotype and genotype. Pick's disease, now accepted as a prototypical FTD, was first recognized early in the last century (Kertesz 1998). As with Alzheimer's disease (AD), this identification was achieved when a German neuropathologist used newly developed vital dyes to identify characteristic inclusions in the brain at autopsy in a few patients with dementia. Distinctive clinical features were difficult to identify from the descriptions of these patients, since they seemed to be similar to AD. Indeed, prominent psychiatric symptoms were the chief features of both Alzheimer's and Pick's original patients (Bick 1994; Kertesz 1998). It became clear that FTD patients characteristically had especially prominent language and behavioral symptoms only after AD had been well characterized and systematic prospective studies of patients with "frontal lobe degeneration of the non-Alzheimer type" performed in Sweden became available in the late 1980s (Brun 1987; Gustafson 1987). Interestingly, even in these early Swedish studies, recognition of patients likely to have predominantly frontal pathology was aided significantly by the imaging of regional cerebral blood flow with xenon (Risberg 1987). Thus, an incompletely characterized proteotype preceded the definition of clinical phenotype.

Our understanding of FTD advanced significantly over the past decade as variations in its proteotype were identified and mutations in the tau gene were found to underlie familial cases (Hutton et al. 1998). Neuropathological classification of FTD changed significantly with the application of immunohistochemistry. Studies have revealed great diversity in pathology, but no consensus has emerged about the best neuropathological classification for FTD. The Alzheimer's Disease Centers, funded by the National Institute on Aging, recently have adopted one scheme through collaborative efforts of the National Alzheimer Coordinating Center (NACC; Table 1). It is noteworthy that this scheme depends solely on pathological observations and sometimes may conflict with clinical findings. For example, the presence or absence of motor neuron disease or parkinsonism in this classification is determined irrespective of findings on neurological examina-

Table 1. NACC pathological classification of FTD

- Pick's disease
- Corticobasal degeneration
- Progressive supranuclear palsy
- Frontotemporal dementia and parkinsonism with tau-positive or argyrophilic inclusions
- Tauopathy, other (e.g., tangle-only dementia and argyrophilic grain dementia)
- FTD with ubiquitin-positive (tau-negative) inclusions (with or without motor neuron disease)
- FTD with no distinctive histopathology (tau-negative, ubiquitin-negative, and no argyrophilic inclusions)
- FTD not otherwise specified (including when immunostaining for ubiquitin and tau not done)

Table 2. Some proposed clinical classification schemes for FTD

The Lund and Manchester Group, 1994
- Frontotemporal dementia (symmetric distribution)
- Progressive aphasia (predominant involvement of language-dominant frontal and temporal lobes)
- Motor neuron disease form
Neary et al. 1998 (the term "frontotemporal lobar degeneration" was applied to all)
- Frontotemporal dementia
- Progressive non-fluent aphasia
- Semantic aphasia and associative agnosia
Edwards-Lee et al. 1997 (subtypes of temporal variants of frontotemporal dementia)
- Left temporal lobe variant
- Right temporal lobe variant
Razani et al. 2001 (subtypes of frontal dementia)
- Left frontal lobe variant
- Right frontal lobe variant
Gorno-Tempini et al. 2004 (subtypes of primary progressive aphasia)
- Nonfluent progressive aphasia
- Semantic dementia
- Logopenic progressive aphasia

tion. Clinical classification of FTD also has evolved over recent years, based upon the relative prominence or absence of presumed characteristic clinical symptoms. Several schemes have been proposed, but none has found general acceptance (Table 2). Definitions offered for these classifications are subjective and open to substantial interpretation. Patients who are not archetypal can be difficult to classify, since category boundaries tend to blur, particularly with dementia progression. The variety of clinical classifications schemes has led to some confusion in the literature and made the diagnosis and management of FTD less accessible to community physicians. Indeed, a recent consensus group concluded that clinical categorization of FTD is unwise and should be abandoned in favor of designating all cases as simply FTD (McKhann et al. 2001).

Unfortunately, thus far it has been impossible to cogently relate clinical and neuropathological phenotypes. Since clinical practice requires diagnosis based upon accessible methods and neuropathological diagnosis has been refined through better identification of specific biochemical abnormalities, neuropathological and clinical schemes have increasingly diverged. Clinical diagnosis has benefited from the improved recognition of affected cognitive domains. Because these cognitive domains reflect anatomic localization, they lend themselves well to neuroimaging, which can more directly evaluate a pattern of selective neuronal injury. Glucose hypometabolism measured by FDG-PET primarily reflects synaptic activity (Mata et al. 1980). Consequently, it is likely to be affected early in disease, even before neuronal loss causes changes in brain volume that might be recognized by structural imaging. To better understand how metabolic phenotypes relate to neuropathological phenotypes, we examined FDG-PET findings in our patients with FTD. The University of Michigan Medical School Review Board for the Institutional Review of Human Subjects Research approved this study.

Subjects

We identified 19 patients with dementia who had an FDG-PET scan at the University of Michigan (between August 1985 and August 1998) and subsequently had an autopsy completed before March 2004 documenting a histopathological diagnosis of FTD other than corticobasal degeneration or progressive supranuclear palsy. For this study, we included only causes of FTD that lacked distinctive motor symptoms, so that metabolic abnormalities would reflect affected cognitive domains. This group of FTD subjects included 11 men and 8 women with a mean age of symptom onset of 60 ± 7 (SD; age range 49 to 81) and a mean age at scan of 65 ± 7 (range 55 to 87; Table 3). They represented a wide range of disease severity and duration at the time of the scan. Nine of the 19 subjects had distinctive histopathology – five had Pick bodies, three had ubiquitin-positive inclusions, and one had tau inclusions. The patient with tau inclusions was a member of a well-described kindred with a (+14) exon 10, splice site mutation of the tau gene (Lynch et al. 1994). FDG-PET scans of these subjects were compared to those of 33 normal elderly subjects (14 women, 19 men; mean age at scan 68 ± 8, age range 58–91).

Image Analysis

FDG-PET data were obtained from archived files. An automated image analysis program, stereotactic surface projection (SSP), was used to display FDG-PET images. SSP summarizes imaging data in three dimensions to aid interpretation of

Table 3. Clinical and neuropathological features of 19 patients with FTD

Case #	Gender	Age at scan	Symptom duration (yrs)	MMSE score	FTD category	Pathology notes
1	Male	61	3	17	FTD NOS	
2	Male	64	5	19	FTD NOS	
3	Male	62	1	20	FTD NOS	
4	Male	59	1	N/A	FTD NOS	See Foster et al. 1992
5	Female	66	1	2	FTD NOS	
6	Male	64	7	25	FTD NOS	
7	Male	64	14	22	FTD NOS	Also plaques and tangles; intermediate probability of AD
8	Male	69	3	0	FTD NOS	
9	Female	87	6	19	FTD NOS	
10	Female	65	7	0	FTD NOS	
11	Male	55	6	27	FTD U+	Motor neuron disease present
12	Male	76	7	8	FTD U+	Motor neuron disease present
13	Female	59	8	13	FTD U+	Motor neuron disease absent
14	Female	64	10	9	FTDP tau+	
15	Male	63	3	20	Pick's disease	GP and SN degeneration also present
16	Female	62	2	24	Pick's disease	
17	Female	61	1	24	Pick's disease	
18	Male	67	8	9	Pick's disease	
19	Female	69	6	23	Pick's disease	

FTDP = frontotemporal dementia with Parkinsonism; GP = globus pallidus; MMSE = Mini-mental Status Examination;

scans (Minoshima et al. 1995). SSP is particularly suited to assessing the cerebral cortex and it is relatively resistant to the affects of atrophy because it selects peak cortical values (Ishii et al. 2001). SSP results are displayed as values of glucose metabolism relative to pons and pixel-by-pixel z-score maps comparing an individual scan with results in the 33 normal controls using a color scale indicating the degree of glucose hypometabolism. Scans were numerically coded and rated

Table 4. Global FDG-PET scan findings in 19 patients with FTD

Case #	FTD category	Overall severity	No. of hypometabolic regions	Hemispheric asymmetry	Scan interpretation notes
1	FTD NOS	Mild	3	R more hypometabolic	
2	FTD NOS	Moderate	4	None	Spread into contiguous inferolateral temporal cortex
3	FTD NOS	Moderate	4	None	Primarily inferior frontal
4	FTD NOS	Moderate	4	None	Primarily superior frontal
5	FTD NOS	Moderate	6	None	Anterior cingulate most affected
6	FTD NOS	Moderate	6	None	Diagnostically ambiguous
7	FTD NOS	Severe	6	R more hypometabolic	
8	FTD NOS	Severe	8	None	
9	FTD NOS	Severe	10	None	
10	FTD NOS	Severe	10	None	Primarily superior frontal
11	FTD U+	Mild	5	R more hypometabolic	Diagnostically ambiguous
12	FTD U+	Moderate	2	None	
13	FTD U+	Severe	8	L more hypometabolic	
14	FTDP tau+	Mild	6	L more hypometabolic	
15	Pick's disease	Moderate	5	L more hypometabolic	diaschiasis; superior > inferior frontal
16	Pick's disease	Moderate	6	R more hypometabolic	Spread into contiguous inferolateral temporal cortex
17	Pick's disease	Moderate	6	R more hypometabolic	
18	Pick's disease	Moderate	7	L more hypometabolic	Inferior > superior frontal
19	Pick's disease	Moderate	7	L more hypometabolic	Spread into contiguous inferolateral temporal cortex

L = left; R = right; other abbreviations as in Table 3.

by a single neurologist (NF) who was experienced in interpretation of FDG-PET scans but was kept unaware of the specific neuropathological classification of FTD. Both metabolic and statistical SSP maps were used to first categorize the severity of scan abnormality as normal, mild, moderate, or severe, based upon both degree and topographic extent of hypometabolism. Next, five specific brain regions in each hemisphere that are most critical to the FDG-PET diagnosis of AD and FTD were rated as normal or abnormal. These areas were the frontal association cortex, anterior temporal cortex, and anterior cingulate cortex, posterior temporoparietal association cortex, and posterior cingulate cortex. Finally, the presence or absence of significant asymmetry in hypometabolism between the left and right cerebral hemispheres was rated, and other notable features in the pattern of glucose hypometabolism were noted.

Results

All scans were abnormal, and sometimes all 10 areas evaluated were affected (Table 5). However, a common theme emerged in which frontal and anterior temporal areas appeared most abnormal, irrespective of how many areas were hypometabolic. Using this as a criterion, diagnosis based upon imaging alone would be unambiguous in all but two cases. Even in these cases, such prominent involvement of anterior temporal and anterior cingulate cortex would be very unusual in AD, where hypometabolism is typically most evident in posterior association areas. In some cases, the superior frontal cortex (Fig. 1) and in other cases the inferior frontal cortex (Fig. 2) appeared most hypometabolic. In several cases, the anterior temporal lobes were affected with relatively little involvement of frontal regions (Fig. 3). Frequently, hypometabolism in the anterior temporal cortex appeared to spread into contiguous areas of the inferolateral temporal cortex, which is commonly affected in AD (Fig. 3). Often, hypometabolism in both frontal and posterior association areas was sufficient to demonstrate the relative sparing of primary motor-sensory cortex (Fig. 2, Table 5, a feature often felt to be characteristic of AD (Foster et al. 1999). Nevertheless, even in these cases with hypometabolism in both frontal and posterior association regions, the abnormalities tended to be more severe in frontal regions, making them distinguishable from scans in patients with AD.

Hemispheric asymmetry was apparent in 10/19 patients (53%) and predominantly affected the right and left hemispheres in equal proportion. Significant asymmetries sometimes were evident even with severe dementia. In one case, there appeared to be crossed cerebellar diaschiasis (Fig. 1), a finding usually observed with frontal strokes or tumors (Miyazawa et al. 2001; Otte et al. 1998). Spearman rank correlations failed to show a significant relationship between either the severity of dementia measured by MMSE or disease duration with the number of hypometabolic regions that appeared hypometabolic. However, patients whose scans were rated as severe tended to have a lower MMSE score (11 ± 10 vs. 17 ± 9) and longer duration of symptoms (8 ±4 vs. 4 ± 3) than those with scans rated either mild or moderate.

Table 5. Regional FDG-PET scan findings in 19 patients with FTD

Case #	FTD category	Frontal		Anterior cingulate		Anterior temporal		Temporo-parietal		Posterior cingulate		Evident sparing of motor-sensory cortex
		R	L	R	L	R	L	R	L	R	L	
1	FTD NOS	-	-	Yes	-	Yes	Yes	-	-	-	-	
2	FTD NOS	-	-	-	-	Yes	Yes	Yes	Yes	-	-	
3	FTD NOS	Yes	Yes	Yes	Yes	-	-	-	-	-	-	
4	FTD NOS	Yes	Yes	Yes	Yes	-	-	-	-	-	-	
5	FTD NOS	Yes	Yes	Yes	Yes	-	-	Yes	Yes	-	-	
6	FTD NOS	-	-	Yes	Yes	Yes	Yes	Yes	Yes	-	-	
7	FTD NOS	Yes	Yes	Yes	Yes	Yes	-	Yes	-	-	-	Only on R
8	FTD NOS	Yes	Yes	Yes	Yes	-	-	Yes	Yes	Yes	Yes	
9	FTD NOS	Yes	Yes	Yes	Yes	Yes	Yes	Yes	Yes	Yes	Yes	Bilateral
10	FTD NOS	Yes	Yes	Yes	Yes	Yes	Yes	Yes	Yes	Yes	Yes	
11	FTD U+	Yes	Yes	-	-	Yes	-	Yes	Yes	-	-	
12	FTD U+	-	-	-	-	Yes	Yes	-	-	-	-	
13	FTD U+	Yes	Yes	Yes	Yes	-	-	Yes	Yes	Yes	Yes	Bilateral
14	FTDP tau+	Yes	Yes	Yes	Yes	Yes	Yes	-	-	-	-	
15	Pick's disease	-	Yes	Yes	Yes	-	Yes	-	-	-	Yes	
16	Pick's disease	Yes	Yes	Yes	Yes	Yes	-	Yes	-	-	-	
17	Pick's disease	Yes	Yes	Yes	Yes	Yes	-	Yes	-	-	-	
18	Pick's disease	Yes	Yes	Yes	Yes	-	Yes	-	Yes	-	Yes	Only on L
19	Pick's disease	-	Yes	Yes	Yes	Yes	Yes	-				

L = left; R = right; other abbreviations as in Table 3.

No single brain region was consistently affected in all patients (Table 5). Frontal and anterior cingulate regions were most often hypometabolic, but anterior temporal lobes sometimes were hypometabolic without involvement of either anterior cingulate or frontal cortex. Although all patients with Pick's disease showed significant hemispheric asymmetry, there was no clear relationship between the

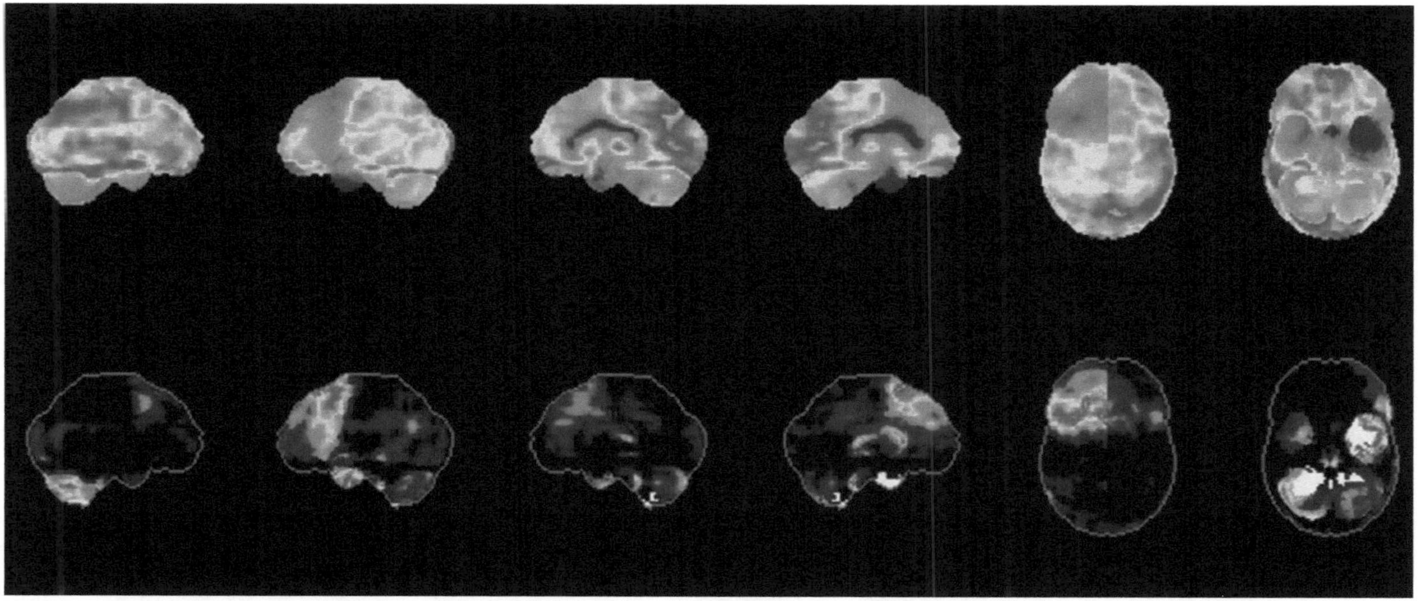

Fig. 1. Stereotactic surface projection (SSP) maps from Case 15, a patient with Pick's disease. SSP maps provide 6 views of the brain – right and left lateral, right and left medial, superior and inferior (in order from left to right on the figure). The top row shows cerebral glucose metabolism relative to the pons. The bottom row shows statistical SSP maps of pixels with significant z-scores compared to 33 elderly normal control subjects. Values in all images are shown in a color scale corresponding to the bar with values red > yellow > green > blue. This scan has significant asymmetry, with the left hemisphere demonstrating greater hypometabolism. Hypometabolism predominantly affects the left superior frontal cortex and right anterior temporal cortex. The anterior cingulate cortex is affected more on the left than the right. In addition, there is hypometabolism in the right cerebellar hemisphere, representing crossed cerebellar diaschisis.

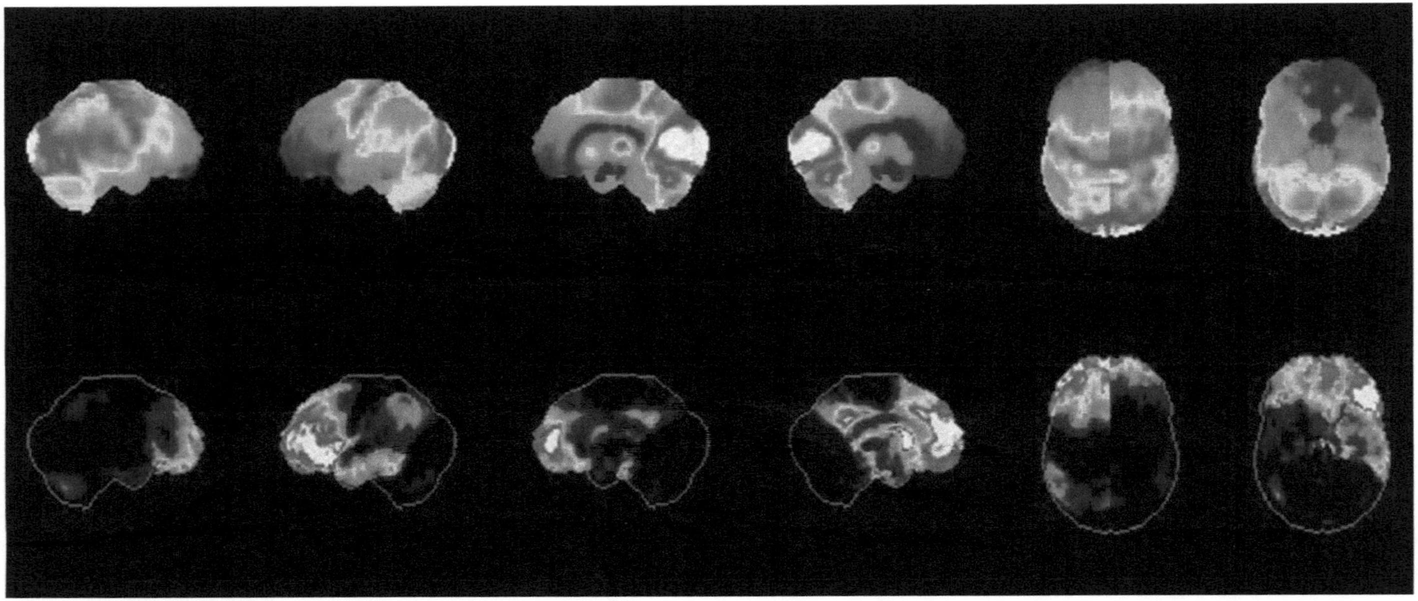

Fig. 2. SSP maps from Case 18, another patient with Pick's disease. These maps are shown in the same format as used in Figure 1. This scan has significant asymmetry, with the left hemisphere demonstrating greater hypometabolism. Hypometabolism affects the inferior more than superior portions of the frontal cortex, as seen in both the lateral and medial projections. In addition to bilateral hypometabolism in the anterior cingulate cortex, the left anterior temporal, left posterior cingulate, and left temporoparietal regions are hypometabolic. This scan is also notable for the evident relative sparing of the sensory-motor cortex in the left hemisphere, which stands out surrounded by affected frontal and posterior association regions as a vertical red band.

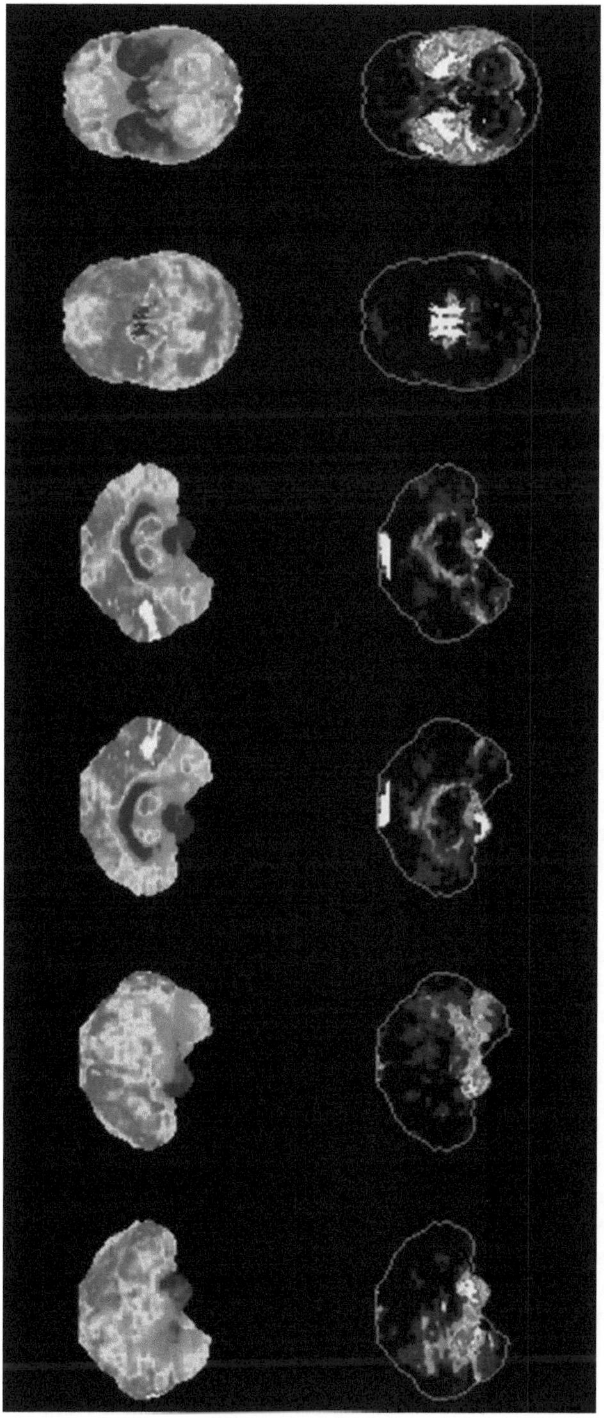

Fig. 3. SSP maps from Case 3, a patient with FTD lacking distinctive histopathology. These maps are shown in the same format as used in Figure 1. This scan shows symmetric hypometabolism affecting primarily anterior temporal cortex, with no significant abnormalities in the frontal cortex or in the anterior cingulate gyrus. While hypometabolism is most pronounced in the anterior temporal cortex, somewhat less severe hypometabolism extends into contiguous regions of the lateral temporal association cortex bilaterally. Without a threedimensional image display, this could be mistakenly interpreted as a predominantly posterior pattern suggesting AD. The very top of the brain was out of the scanner's field of view, accounting for the white regions at the top of the medial and superior views, which are technical artifacts.

pattern of glucose hypometabolism and the neuropathologic classification of patients. We were also unable to identify any particular imaging characteristic that distinguished the patient with a known tau mutation from other FTD cases.

Discussion

There is a dependable pattern of glucose hypometabolism in FTD that is consistent with the known distribution of neuronal and synaptic loss in frontal and anterior temporal brain regions. This finding suggests that the localization of metabolic abnormalities may aid in diagnosis, particularly since most patients with FTD also meet that clinical criteria for AD (Varma et al. 1999). However, there is much variation within this general pattern of glucose hypometabolism. Further research is needed to understand the environmental and genetic factors that modify disease expression and explain these variations. We were unable to find evidence that disease proteotype represented by current neuropathological classification or the presence of a tau mutation explains the variability to regional glucose metabolism.

Although we cannot yet predict or explain variations in the pattern of glucose hypometabolism, these changes appear to correspond quite well to variations in clinical symptoms and phenotype (Edwards-Lee et al. 1997; Miller et al. 1997). To the extent that identifying clinical phenotype is useful, FDG-PET may provide a more objective method to support clinical classification of patients with FTD. Adding information about cerebral metabolism would be particularly useful in expanding our study of FTD beyond only those patients with classical, prototypical features and early in the illness when only a few of the typical symptoms of FTD are evident (Mendez and Perryman 2002). Even patients with FTD due to the same genetic mutations can have remarkably diverse clinical presentations and pathology (Nasreddine et al. 1999; Zhukareva et al. 2003), and a better understanding of the location and extent of brain pathology could provide important information to guide therapy.

Our findings also have implications for clinical diagnosis. Although the onset of FTD is relatively more common in individuals less than 60 years of age, our experience demonstrates that early age of onset of dementia should not be a diagnostic criterion for FTD. It is likely that many patients with late onset FTD are misdiagnosed as having AD or never identified. We found that metabolic changes in FTD are not limited to frontal and anterior temporal regions; posterior association cortex can also be affected. Consequently, hypometabolism in posterior cortical regions and relative sparing of the primary motor and sensory cortex should not be misinterpreted as evidence of AD when frontal or anterior temporal areas are more affected. It should be noted that quantitative neuropathological assessment has demonstrated predominant frontal distribution of typical amyloid plaques and neurofibrillary tangles in three patients with correspondingly prominent frontal symptoms (Johnson et al. 1999). Although no molecular imaging was performed in these patients, they illustrate further that linkage of proteotype and localization of pathology is not absolute. More studies are needed to understand how metabolic abnormalities evolve in FTD and what the pathologic explanation is for both the selective neuronal vulnerability characteristic of this disease and how more pervasive metabolic deficits can develop in less susceptible brain regions.

Acknowledgments

This work was supported in part by grants from the National Institute on Aging of the National Institutes of Health P50-AG08671 (Michigan Alzheimer's Disease Research Center) and RO1-AG22394 and a collaborative pilot grant from the National Alzheimer Coordinating Center, U01-AG16976

References

Bick KL (1994) The early story of Alzheimer disease In: Terry RD, Katzman R, Bick KL (eds) Alzheimer disease. Raven Press, New York, pp 1–8

Brun A (1987) Frontal lobe degeneration of non-Alzheimer type. I. Neuropathology. Arch Gerontol Geriatr 6:193–208

Edwards-Lee T, Miller BL, Benson DF, Cummings JL, Russell GL, Boone K, Mena I.(1997) The temporal variant of frontotemporal dementia. Brain 120:1027–1040

Foster NL, Gilman S, Berent S, Sima AAF, D'Amato C, Koeppe RA, Hicks SP. (1992) Progressive subcortical gliosis and progressive supranuclear palsy can have similar clinical and PET abnormalities. J Neurol Neurosurg Psychiat 55:707–713

Foster NL, Minoshima S, Kuhl DE (1999) Brain imaging in Alzheimer disease. In: Terry RD, Katzman R, Bick KL, Sisodia SS (eds) Alzheimer Disease. 2nd ed. Raven Press, New York, pp 67–93

Gorno-Tempini ML, Dronkers NF, Rankin KP, Ogar JM, Phengrasamy L, Rosen HJ, Johnson JK, Weiner MW, Miller BL (2004) Cognition and anatomy in three variants of primary progressive aphasia. Ann Neurol 55:335–346

Gustafson L (1987) Frontal lobe degeneration of non-Alzheimer type II. Clinical picture and differential diagnosis. Arch Gerontol Geriatr 6:209–223

Hutton M, Lendon CL, Rizzu P, Baker M, Froelich S, Houlden H, Pickering-Brown S, Chakraverty S, Isaacs A, Grover A, Hackett J, Adamson J, Lincoln S, Dickson D, Davies P, Petersen RC, Stevens M, de Graaff E, Wauters E, van Baren J, Hillebrand M, Joosse M, Kwon JM, Nowotny P, Che LK, Norton J, Morris JC, Reed LA, Trojanowski J, Basun H, Lannfelt L, Neystat M, Fahn S, Dark F, Tannenberg T, Dodd PR, Hayward N, Kwok JBJ, Schofield PR, Andreadis A, Snowden J, Craufurd D, Neary D, Owen F, Oostra BA, Hardy J, Goate A, van Swieten J, Mann D, Lynch T, Heutink P (1998) Association of missense and 5'-splice-site mutations in tau with the inherited dementia FTDP-17. Nature 393:702–705

Ishii K, Willoch F, Minoshima S, Drzezga A, Ficaro EP, Cross DJ, Kuhl DE, Schaiger M (2001) Statistical brain mapping of 18F-FDG PET in Alzheimer's disease: validation of anatomic standardization for atrophied brains. J Nucl Med 42:548–557

Johnson JK, Head E, Kim R, Starr A, Cotman CW (1999) Clinical and pathological evidence for a frontal variant of Alzheimer disease. Arch Neurol 56:1233–1239

Kertesz A (1998) Arnold Pick: A historical introduction. In: Kertesz A, Munoz DG (eds) Pick's disease and Pick complex. Wiley-Liss, New York, pp 13–21

The Lund and Manchester Groups (1994) Clinical and neuropathological criteria for frontotemporal dementia.. J Neurol Neurosurg Psychiat 57:416–418

Lynch T, Sano M, Marder KS, Bell KL, Foster NL, Defendini RF, Sima AAF, Keohane C, Nygaard TG, Fahn S, Mayeux R, Rowland LP, Wilhelmsen KC (1994) Clinical characteristics of a family with chromosome 17-linked disinhibition-dementia-parkinsonism-amyotrophy complex. Neurology 44:1878–1884

Mata M, Fink DJ, Gainer H, Smith CB, Davidsen L, Savaki H, Schwartz WJ, Sokoloff L (1980) Activity-dependent energy metabolism in rat posterior pituitary primarily reflects sodium pump activity. J Neurochem 34:213–215

McKhann GM, Albert MS, Grossman M, Miller B, Dickson D, Trojanowski JQ (2001) Clinical and pathological diagnosis of frontotemporal dementia: report of the Work Group on Frontotemporal Dementia and Pick's Disease. Arch Neurol 58:1803–1809

Mendez MF, Perryman KM (2002) Neuropsychiatric features of frontotemporal dementia: evaluation of consensus criteria and review. J Neuropsychiat Clin Neurosci 14:424–429.

Miller BL, Ikonte C, Ponton M, Levy M, Boone K, Darby A, Berman N, Mena I, Cummings JL (1997) A study of the Lund-Manchester research criteria for frontotemporal dementia: clinical and single-photon emission CT correlations. Neurology 48:937–942

Minoshima S, Frey KA, Koeppe RA, Foster NL, Kuhl DE (1995) A diagnostic approach in Alzheimer's disease using 3-dimensional sterotactic surface projections of fluorine-18-FDG PET. J Nucl Med 36:1238–1248

Miyazawa N, Toyama K, Arbab AS, Koizumi K, Arai T, Nukui H (2001) Evaluation of crossed cerebellar diaschisis in 30 patients with major cerebral artery occlusion by means of quantitative I-123 IMP SPECT. Ann Nucl Med 15:513–519

Nasreddine ZS, Loginov M, Clark LN, Lamarche J, Miller BL, Lamontagne A, Zhukareva V, Lee VM, Wilhelmsen KC, Geschwind DH (1999) From genotype to phenotype: a clinical pathological, and biochemical investigation of frontotemporal dementia and parkinsonism (FTDP-17) caused by the P301L tau mutation. Ann Neurol 45:704–715

Neary D, Snowden JS, Gustafson L, Passant U, Stuss D, Black S, Freedman M, Kertesz A, Robert PH, Albert M, Boone K, Miller BL, Cummings J, Benson DF (1998) Frontotemporal lobar degeneration: a consensus on clinical diagnostic criteria. Neurology 51:1546–1554

Otte A, Roelcke U, von Ammon K, Hausmann O, Maguire RP, Missimer J, Muller-Brand J, radu EW, Leenders KL (1998) Crossed cerebellar diaschisis and brain tumor biochemistry studied with positron emission tomography, [18F]fluorodeoxyglucose and [11C]methionine. J Neurol Sci 156:73–77

Razani J, Boone KB, Miller BL, Lee A, Sherman D (2001) Neuropsychological performance of right- and left-frontotemporal dementia compared to Alzheimer's disease. J Int Neuropsychol Soc 7:468–480

Risberg J (1987) Frontal lobe degeneration of non-Alzheimer type III. Regional cerebral blood flow. Arch Gerontol Geriatr 6:225–233

Varma AR, Snowden JS, Lloyd JJ, Talbot PR, Mann DM, Neary D (1999) Evaluation of the NINCDS-ADRDA criteria in the differentiation of Alzheimer's disease and frontotemporal dementia. J Neurol Neurosurg Psychiat 66:184–188

Zhukareva V, Sundarraj S, Mann D, Sjogren M, Blenow K, Clark CM, McKeel DW, Goate A, Lippa CF, Vonsattel JP, Growdon JH, Trojanowski J, Lee VM (2003) Selective reduction of soluble tau proteins in sporadic and familial frontotemporal dementias: an international follow-up study. Acta Neuropathol (Berl) 105:469–476

Phenotype/genotype correlations in Parkinson's disease

Alexis Brice[1,2,*], *Ebba Lohmann*[1], *Pablo Ibanez*[1], *Magali Periquet*[1], *Sophie Laine*[1], *Béatrice Debarges*[1], *Suzanne Lesage*[1] *and Alexandra Dürr*[1,2]

Introduction

Idiopathic Parkinson's disease (IPD) is a common neurodegenerative condition characterized clinically by three major features: resting tremor, bradykinesia and rigidity. It is caused by the selective degeneration of dopaminergic neurones in the substantia nigra pars compacta, resulting in severe dopamine depletion in the striatum to which it projects. The pathognomonic hallmarks of the disease are intracytoplasmic, ubiquitine-positive, eosinophilic inclusions in surviving brainstem neurones that contain large amounts of α-synuclein. Therefore, a definitive diagnosis of Parkinson's disease (PD) is only possible after post-mortem examination. Because of these inclusions, IPD has been classified among the group of so-called α-synucleinopathies, which also comprises dementia with Lewy bodies (DLB) and multisystem atrophy (Goedert and Spillantini 1998). However, in the light of recent genetic studies, the relevance of these inclusions for diagnostic purposes has been brought into question.

Although the recognition that some forms of PD have a Mendelian pattern of inheritance is not new, it is only in the last decade that responsible loci/genes have been identified (Lansbury and Brice 2002 ; Bonifati et al. 2003; Hardy et al. 2003). So far, eight loci have been implicated in monogenic forms of PD and five of the causative genes have been identified (Table 1). Three loci account for some of the autosomal recessive forms of parkinsonism with early onset: PARK2 (Parkin gene; Kitada et al. 1998), PARK6 (Pink1 gene; Valente et al. 2004) and PARK7 (DJ-1 gene; Bonifati et al. 2003). In autosomal dominant forms, three genes have been identified: PARK1/4 (α-synuclein gene; Polymeropoulos et al. 1997), PARK5 (ubiquitin carboxyterminal-hydrolase L1 gene; Leroy et al. 1998) and NR4A2 (Nurr1 gene; Le et al. 2003). However, since mutations in the latter two have not been found in any subsequent studies, the causative role of the variants initially identified remains doubtful from a genetic point of view. Three other loci have been mapped to chromosomes 2p13 (PARK2; Gasser et al. 1998), 12p11 (PARK8; Funayama et al. 2002) and 1p32 (PARK10; Hicks et al. 2002) under established or presumable autosomal dominant models. In addition, there is evidence for several genetic susceptibility loci identified through genome linkage screens in large cohorts of sib-pairs and nuclear families (Martinez et al. 2004).

[1] INSERM Unit 679; [2] Département de Génétique, Cytogénétique et Embryologie; Hôpital de la Pitié-Salpêtrière, AP-HP ; 47 boulevard de l'Hôpital ;75651 PARIS cedex 13, France
* INSERM Unit 679 and Département de Génétique, Cytogénétique et Embryologie; Hôpital de la Pitié-Salpêtrière, AP-HP; 47 boulevard de l'Hôpital; 75651 PARIS cedex 13, E-mail: brice@ccr.jussieu.fr

Cummings et al.
Genotype – Protéotype – Phenotype
Relationships in Neurodegenerative Diseases

Table 1. Monogenic forms of Parkinson's disease

Designation	Locus	Gene	Trans-mission	Mean age at onset (years)	Progression	Lewy bodies
PARK1	4q21-33	α-synuclein point mutations	AD	Variable	Severe	+
PARK2	6q25-2.27	Parkin	AR	Early	Very slow	- (except one case)
PARK3	2p13	?	AD	Late	Slow	+
PARK4	4q21	α-synuclein rearrange-ments	AD	Variable	Variable	+
PARK5	4p14	UCH-L1	Probable AD	50	?	ND
PARK6	1p35-36	Pink1	AR	Early	Slow	ND
PARK7	1p36	DJ-1	AR	Early	Slow	ND
PARK8	12p11.2-q13.1	?	AD	Late	?	Variable

AD, autosomal dominant; AR, autosomal recessive; ND, not determined.

As a result of the identification of monogenic forms of PD, we can now gain insight into the mechanisms of the disease, but also improve genetic counselling through molecular testing and deepen our understanding of the genetic aspects of the disease, in particular through the establishment of genotype/phenotype correlations. Given the limited number of families with monogenic forms of PD, probably no more than a few percent of patients, clear-cut phenotypic differences can only be established for the most frequent genes: α-synuclein and Parkin. Therefore, only the characteristics of parkinsonism resulting from alterations in these two genes will be presented in detail.

α-synuclein mutations in PD

The A53T mutation in the α-synuclein gene was the first to be identified in PD (Polymeropoulos et al. 1997). It was followed by two other missense mutations: A30P (Kruger et al. 1998) and E46K (Zarranz et al. 2004). The A53T mutation initially identified in an Italo-American pedigree, the Contursi family (Polymeropoulos et al. 1997), is rare but has since been found in several pedigrees living in Greece, or of Greek origin (Bostantjopoulou et al. 2001; Papapetropoulos et al. 2001; Spira et al. 2001), probably due to a founder effect (Athanassiadou et al. 1999). The characteristics of the disease in the Contursi kindred are those of typical PD, with an early age at onset (a mean of 45 years) and rapid progression of the disease to death (a mean of nine years), as well as almost complete penetrance in gene carriers (Golbe et al. 1996). In addition, cognitive decline and hallucinations have been noted in several patients. Analyses of additional kindreds with the same mutation have confirmed the tendency for early onset compared to IPD but have extended the phenotype. In an Australian kindred of Greek origin, Spira et al. (2001) noted hypoventilation of central origin, myoclonus, postural hypotension and urinary incontinence. The clinical features of the few patients with the A30P mutation remain within the spectrum of PD (Kruger et al. 1998). In contrast, the phenotype is quite different in patients with the E46K mutation (Zarrantz et al. 2004): the age at onset is usually later (between age 50 and 65), with frequent dementia, hallucinations and fluctuations of consciousness reminiscent of DLB. These additional signs are consistent with the neuropathological data showing that the brain lesions are more widespread than would be expected in IPD. The initial histopathological study of A53T cases revealed neuronal degeneration, gliosis and Lewy bodies in the substantia nigra and locus coeruleus as well as some cortical Lewy bodies compatible with PD. In the Australian family, in addition to features of PD, there were neuritic changes in all cortical areas but predominantly in medial temporal regions, associated with tissue vacuolisation and marked gliosis in the basal ganglia, but no cortical Lewy bodies. The only case with the E46K mutation that has been studied presents, in addition to the pathological features of PD, numerous Lewy bodies and Lewy neurites in the cerebral cortex and more numerous still in the para-hippocampus and amygdala (Zarrantz et al. 2004). These features are typical of DLB.

Our view of α-synuclein mutations has recently been enlarged by the identification of families with rearrangements of the 4q21 region including this gene. First, a triplication of the α-synuclein and several neighbouring genes was detected in the Spellman-Muenter kindred with autosomal dominant PD (Singleton et al 2003) followed by a Swedish–American kindred (Farrer et al. 2004). More recently, duplications of the region encompassing the α-synuclein gene were also detected in French families with autosomal dominant PD (Ibanez et al. 2004; Chartier-Harlin et al. 2004). Interestingly, these rearrangements appear to be more frequent than the point mutations. In the French population, α-synuclein gene duplications were found in one of the nine large families tested by Chartier-Harlin et al. (2004) and in two of the 119 families with PD with presumably autosomal dominant transmission (Ibanez et al. 2004), whereas no point mutations were detected in this population. Interestingly, the rearrangements of the α-synuclein gene seem to have occurred independently and differ in size (Farrer et al 2004; Ibanez et al. 2004). However, in all cases, the entire α-synuclein gene is involved, and it has been shown that the amount of the corresponding mRNA and protein in blood is doubled in individuals with triplications. Furthermore, in the brain of a patient with a triplication, there is an increase in aggregated forms of the protein (Miller et al. 2004). Perhaps most interesting is the clear effect of gene dosage detected at the clinical level. Patients with duplications have a phenotype similar to IPD but with autosomal dominant transmission (Chartier-Harlin et al. 2004; Ibanez et al. 2004). Their age at onset ranges from 39 to 65, and the mean duration of the disease to death is 17±7.2 years. Cognitive decline is not prominent and always appears late in the disease. No cases with duplication have as yet been autopsied. Triplications are associated with a phenotype reminiscent of DLB with autosomal dominant transmission and early onset. In the Spellman-Muenter family, known as the Iowa kindred, the mean age at onset is 35±8.6, and duration to death is 8.4±3.7 years (Muenter et al. 1998). In addition, patients present with atypical signs such as hallucinations, early and severe dementia and postural hypotension. Neuropathological examination reveals abundant α-synuclein pathology in the brainstem and Lewy neurites throughout the neocortex (Muenter et al. 1998). A striking feature in both kindreds with α-synuclein gene triplications is severe neuronal loss in the CA2/3 region of the hippocampus, also seen in patients with point mutations in this gene.

In summary, point mutations and multiplications of the α-synuclein gene, both of which are associated with autosomal dominant transmission, have different consequences. A classical PD phenotype is caused by duplications of the α-synuclein gene, resulting in three functional copies instead of two. A severe phenotype with dopa-responsive parkinsonism and additional atypical features resembling DLB is associated with missense mutations (particularly the A53T and E47K) and with triplications of the α-synuclein gene. However, triplications in which four copies of the α-synuclein gene are expressed instead of two seem to result in an even earlier age at onset. Both missense mutations and triplications produce widespread lesions, as in DLB, but with severe neuronal loss in the hippocampus.

Parkin gene mutations

Autosomal recessive juvenile parkinsonism (AR-JP) is a clinical entity that was first distinguished in Japan (Yamamura et al. 1973). This entity is characterized by early onset (before the age of 40 but often before 20), sleep benefit, dystonia (mainly in the feet), resting tremor that is less frequent than in IPD, brisk reflexes, and a good response to levodopa. Dementia and autonomic failure are not part of the phenotype. In addition to this unique clinical phenotype, these patients have selective loss of pigmented neurones in the substantia nigra and locus coeruleus that is not associated with Lewy bodies (Takahashi et al. 1994). Once recognized, families with this disorder were collected and the responsible gene was mapped to chromosome 6q25.2-q27 (Matsumine et al. 1997). Linkage studies in families with early onset autosomal recessive (EOAR) PD then showed that the disease was also present in European and North African populations (Tassin et al. 1998; Jones et al. 1998). These results were confirmed when the responsible gene was identified. In Japanese families with AR-JP, Kitada et al. (1998) detected several deletions in a new gene encoding a 465 amino acid protein that they called Parkin. This finding was the starting point for numerous studies aimed at determining the frequency and spectrum of parkin mutations as well as the associated clinical and biochemical phenotype.

The spectrum of mutations is very varied, including point mutations of different types as well as rearrangements with deletions and multiplications of one or more exons (Fig. 1; Kitada et al. 1998; Lucking et al. 1998; Hattori et al. 1998; Abbas

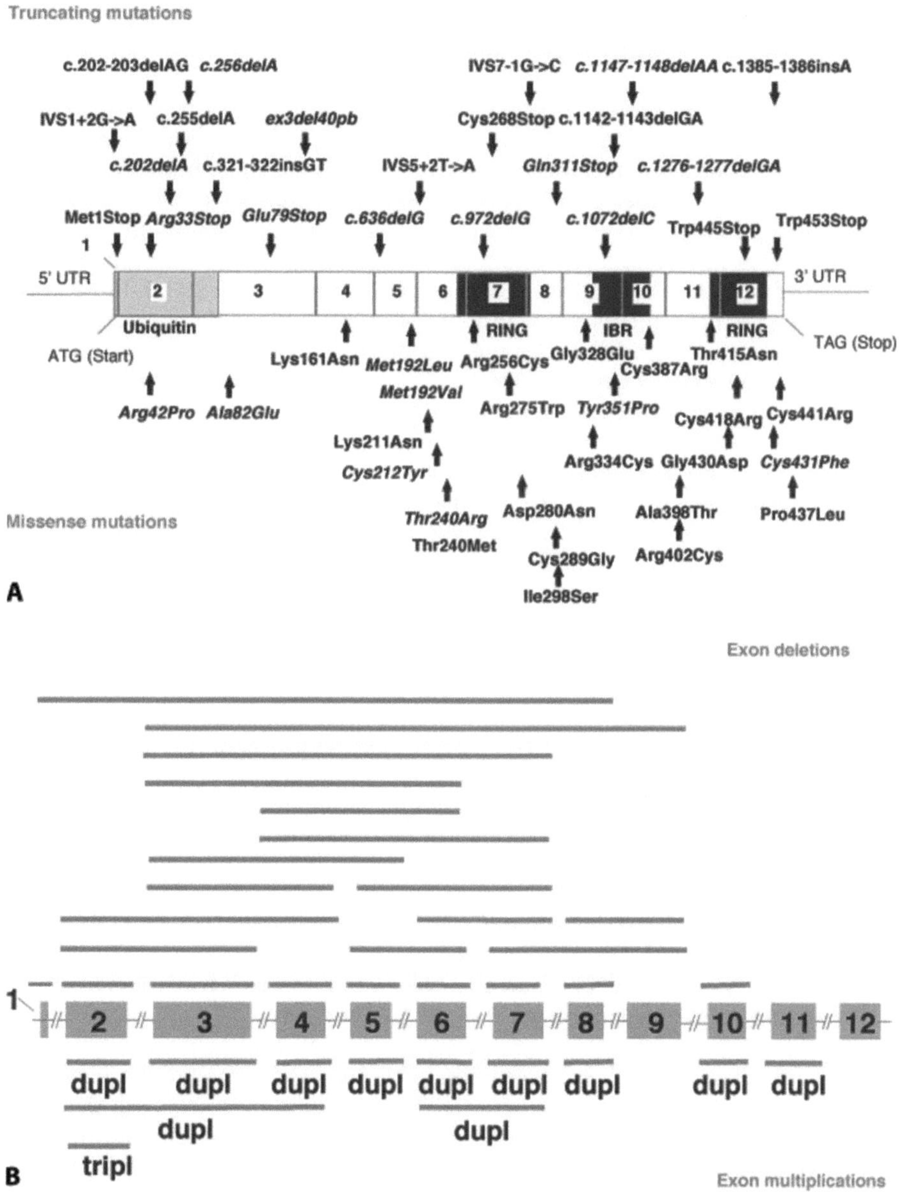

Fig. 1. Point mutations and exon rearrangements in the Parkin gene. **A** Point mutations, which are predicted to produce an absent or a truncated protein, are indicated above, and missense mutations appear below the schematic representation of the Parkin cDNA. **B** The size and the position of deletions and multiplications [duplications (dupl) and triplications (tripl)]) are represented by lines above and below the schematic representation of the Parkin gene, respectively. The functional domains of Parkin are indicated below the schematic representation in A. Ubiquitin = ubiquitin-like domain; RING-IBR-RING = RING-in between Ring-RING; UTR = untranslated region.

et al. 1999; Klein et al. 2000; Periquet et al. 2001; Hedrich et al. 2002; West et al. 2002; Periquet et al. 2003; Rawal et al. 2003). More than a hundred mutations have already been identified. They occur in patients in various combinations of point mutations or small or large rearrangements. Many patients appear to carry only one detectable mutation. The significance of this finding is still debated. It is suspected that the high frequency of exon rearrangements in the Parkin gene is due to its size, 1350 Kb, the second largest in the human genome, compared to the small number of coding exons (12). Interestingly, several point mutations in families from different countries are associated with the same haplotype, suggesting a founder effect (Periquet et al. 2001; Lincoln et al. 2003). In contrast, most exon rearrangements are associated with different haplotypes even in the same population, suggesting that they occurred independently (Periquet et al. 2001). Nevertheless, because of the extreme variety of different types of mutations in the Parkin gene, screening for the mutations is complex, and a combination of sequencing and dosage experiments is needed for accurate detection (Lucking and Brice 2003; Hedrich et al. 2001).

That the frequency of Parkin gene mutations is very high in early onset (EO) PD was a major finding. Several studies analysed large series of patients with either familial or apparently sporadic EO parkinsonism (Hattori et al. 1998; Abbas et al. 1999; Klein et al. 2000; Periquet et al. 2001; Hedrich et al. 2002; West et al. 2002; Periquet et al. 2003; Rawal et al. 2003). In autosomal recessive (AR) forms, parkin mutations represented over 50% of the cases with onset up to age 35 (Periquet et al. 2003). Therefore, the majority of EO cases with AR transmission, defined by an age at onset below 40 years, can be accounted for by Parkin mutations. These results demonstrate that other causative genes for EOAR parkinsonism are less frequently involved. Most surprising was the frequency with which Parkin was found to be involved in EO cases without family histories (sporadic cases). Periquet et al. (2003) reported that parkin mutations accounted for at least 15% of 246 such cases analysed by sequencing and gene dosage. However, the relative frequency of Parkin cases varies greatly with age at onset. It is approximately 2/3 before the age of 20, 1/4 between 20 and 29 and less that 1/10 after 30. Fewer studies have dealt with late onset cases (Oliveri et al. 2001; Kann 20002; Oliveira et al. 2003; Klein et al. 2003; Foroud et al. 2003), but they are clearly less frequent, representing only a small percentage of the patients.

Because of the number of parkin cases found on all continents, the major features of the disease, as well as the range of phenotypical manifestations, have been well characterized. The usual features are EO typical parkinsonism with slow clinical course, good or excellent response to low doses of levodopa with frequent treatment-induced dyskinesias and the absence of dementia (Lucking et al. 2000; Khan et al. 2003). Other frequent signs that occur in less than 50% of the cases are foot dystonia, brisk reflexes, sleep benefit and psychiatric or behavioural disorders. Two studies have compared the frequency of clinical signs in parkin cases and cases recruited according to the same clinical criteria but without mutations in this gene (Lücking et al. 2000; Lohmann et al. 2003). Although the range of ages at onset was similar in both groups, the mean age at onset was significantly earlier in Parkin cases (31.4 ± 11.9 versus 38.1 ± 11.2; $p < 0.001$). Dystonia, symmetry at onset and brisk reflexes were significantly more frequent in parkin cases than in the others. Despite longer treatment, the daily dose of Levodopa was significantly lower in the parkin group. However, there was no individual sign or symptom that distinguished parkin carriers from non-carriers with EO parkinsonism, and there is no effect of gender on the phenotype. Interestingly, dystonia at onset was not a specific sign of parkin carriers but rather was associated with young onset parkinsonism regardless of the cause. The spectrum of the phenotype is, therefore, large, with atypical presentations or additional signs including late-onset PD (Lucking et al. 2000; Klein et al. 2000; Nichols et al. 2002; Oliveira et al. 2003), phenotypes resembling Dopa-responsive dystonia (Tassin et al. 2000; Khan et al. 2003), hemiparkinsonism-hemiatrophy (Pramstaller et al. 2002), forms with cerebellar ataxia (van de Warrenburg et al. 2001; Kuroda et al. 2001; Periquet et al. 2003), pyramidal tract dysfunction (Kuroda et al. 2001), peripheral neuropathy (Tassin et al. 1998), tremor mainly during orthostatism (Rawal et al. 2003) and white matter abnormalities on brain MRI (Lohmann et al. 2003). All of these features extend the Parkin phenotype and are important to take into account when molecular testing is requested.

Several studies also attempted to establish phenotypic correlations with the nature of the mutations or their number (Lohmann et al. 2003; Foroud et al. 2003; Oliveira et al. 2003). The nature of the mutation does not seem to account for much of the phenotypical variability since patients with the same two mutations may have a 27-year difference in the age at onset, including both early and late onset cases (Lohmann et al. 2003). However, the same group has shown that missense mutations located within known functional domain of the protein result in a significantly earlier age at onset than other missense mutations. It was consistently observed that a significant though variable proportion of patients, particularly among those without family histories, carried only one detectable parkin gene mutation. This finding is surprising since the two alleles are expected to carry mutations in AR disorders. The second mutation may have escaped detection because of an atypical mutational mechanism (i.e., inversion) that has not been explored with currently available techniques, or because of the size of the gene that includes large unexplored regions, such as the promoter or the introns, that may have important regulatory functions. Alternatively, it has been postulated that some parkin mutations could be dominant or constitute risk factors for late onset PD (West et al. 2002; Farrer et al. 2002; Foroud et al. 2003; Oliveira et al. 2003). In several series of patients, there was a significant trend towards later onset in cases carrying only one allele with a parkin mutation compared to those with two mutations (Lohmann et al. 2003; Hedrich et al. 2002; Foroud et al. 2003; Oliveira et al. 2003). However, the causative role of single parkin mutations remains to be confirmed by functional data.

Neuropathological examination of parkin cases, as mentioned before, reveals lesions that are strikingly different from IPD. So far, at least five cases with AR-JP or proven parkin mutations have been described (Takahashi et al. 1994; Mori et al. 1998; Hayashi et al. 2000; van de Warrenburg et al. 2001; Farrer et al. 2001; Gouider-Khouja et al. 2003). All but one case with a family history of autosomal dominant parkinsonism (Farrer et al. 2001) had no Lewy bodies but severe generalised loss of dopaminergic neurones in the substantia nigra pars compacta. Other abnormalities were variable. These included additional involvement of the substantia nigra pars reticulata (Hayashi et al. 2000), neurofibrillary tangles and argyrophilic astrocytes in the cerebral cortex and brainstem nuclei (Mori et al. 1998), and neuronal loss in parts of the spinocerebellar systems (van de Warrenburg et al. 2001). A case with a single parkin mutation had tau aggregates as in progressive supranuclear palsy (Morales et al. 2002). These observations suggest that the mechanism by which the loss of Parkin function due to mutations causes parkinsonism with degeneration of dopaminergic neurones might be different from the mechanisms underlying IPD, which is considered to be a synucleinopathy. However, although these are major differences, they have few consequences at the clinical level, where no individual features distinguish parkin disease from IPD. Whether parkin pathology is relevant to IPD remains, therefore, an open question. There is growing evidence, however, that there is a functional link between the α-synuclein and parkin proteins.

Acknowledgments

We thank Merle Ruberg (PhD) for critical reading of the manuscript, the members of the French Parkinson's Disease Genetic Study Group and of the DNA and Cell Bank of IFR 70. This work was supported by grants from the National Institutes of Health (grant NS41723-01A1), INSERM and the French Ministry of Research and Technology "Cohortes et Collections 2001" (contract 4CH03G).

References

Abbas N, Lücking CB, Ricard S, Durr A, Bonifati V, De Michele G, Bouley S, Vaughan JR, Gasser T, Marconi R, Broussolle E, Brefel-Courbon C, Harhangi BS, Oostra BA, Fabrizio E, Bohme GA, Pradier L, Wood NW, Filla A, Meco G, Denefle P, Agid Y, Brice A (1999) A wide variety of mutations in the parkin gene are responsible for autosomal recessive parkinsonism in Europe. French Parkinson's Disease Genetics Study Group and the European Consortium on Genetic Susceptibility in Parkinson's Disease. Human Mol Genet 8:567-574.

Athanassiadou A, Voutsinas G, Psiouri L, Leroy E, Polymeropoulos MH, Ilias A, Maniatis GM, Papapetropoulos T (1999) Genetic analysis of families with Parkinson disease that carry the Ala53Thr mutation in the gene encoding alpha-synuclein. Am J Human Genet 65:555-558.

Bonifati V, Rizzu P, Van Baren M.J, Cshaap O, Breedveld GJ, Krieger E, Dekker MCJ, Squitieri F, Ibanez P, Joosse M, Van Dongen JW, Vanacore N, Van Swieten JC, Brice A, Meco G, Van Duijn CM, Oostra BA, Heutink P (2003) Mutations in the DJ-1 gene associated with autosomal recessive early onset parkinsonism. Science 299:256-259.

Bostantjopoulou S, Katsarou Z, Papadimitriou A, Veletza V, Hatzigeorgiou G, Lees A (2001) Clinical features of parkinsonian patients with the alpha-synuclein (G209A) mutation. Mov Disord 16:1007-1013.

Chartier-Harlin MC, Kachergus J, Roumier C, Mouroux V, Douay X, Lincoln S, Levecque C, Larvor L, Andrieux J, Hulihan M, Waucquier N, Defebvre L, Amouyel P, Farrer M, Destée A (2004) Alpha-synuclein locus duplication causes familial Parkinson s disease. Lancet 364:1167-1169

Farrer M, Chan P, Chen R, Tan L, Lincoln S, Hernandez D, Forno L, Gwinn-Hardy K, Petrucelli L, Hussey J, Singleton A, Tanner C, Hardy J, Langston JW (2001) Lewy bodies and parkinsonism in families with parkin mutations. Ann Neurol 50:293-300.

Farrer M, Kachergus J, Forno L, Lincoln S, Wang DS, Hulihan M, Maraganore D, Gwinn-Hardy K, Wszolek Z, Dickson D, Langston JW (2004) Comparison of kindreds with parkinsonism and alpha-synuclein genomic multiplications. Ann Neurol 55:174-179.
Foroud T, Uniacke SK, Liu L, Pankratz N, Rudolph A, Halter C, Shults C, Marder K, Conneally PM, Nichols WC; Parkinson Study Group (2003) Heterozygosity for a mutation in the parkin gene leads to later onset Parkinson disease. Neurology 60:796-801.
Funayama M, Hasegawa K, Kowa H, Saito M, Tsuji S, Obata F (2002) A new locus for Parkinson's disease (PARK8) maps to chromosome 12p11.2-q13.1. Ann Neurol 51:296-301.
Gasser T, Muller-Myhsok B, Wszolek ZK, Oehlmann R, Calne DB, Bonifati V, Bereznai B, Fabrizio E, Vieregge P, Horstmann RD (1998) A susceptibility locus for Parkinson's disease maps to chromosome 2p13. Nature Genet 18:262-265.
Goedert M, Spillantini MG (1998) Lewy body diseases and multiple system atrophy as alpha-synucleinopathies. Mol Psychiatr 3:462-465.
Golbe LI, Di Iorio G, Sanges G, Lazzarini AM, La Sala S, Bonavita V, Duvoisin RC (1996) Clinical genetic analysis of Parkinson's disease in the Contursi kindred. Ann Neurol 40:767-775.
Gouider-Khouja N, Larnaout A, Amouri R, Sfar S, Belal S, Ben Hamida C, Ben Hamida M, Hattori N, Mizuno Y, Hentati F (2003) Autosomal recessive parkinsonism linked to parkin gene in a Tunisian family. Clinical, genetic and pathological study. Parkinsonism Relat Disord 9:247-251.
Hardy J, Cookson MR, Singleton A (2003) Genes and parkinsonism. Lancet Neurol 2:221-228.
Hattori N, Kitada T, Matsumine H, Asakawa S, Yamamura Y, Yoshino H, Kobayashi T, Yokochi M, Wang M, Yoritaka A, Kondo T, Kuzuhara S, Nakamura S, Shimizu N, Mizuno Y (1998)Molecular genetic analysis of a novel Parkin gene in Japanese families with autosomal recessive juvenile parkinsonism: evidence for variable homozygous deletions in the Parkin gene in affected individuals. Ann Neurol 44:935-941.
Hayashi S, Wakabayashi K, Ishikawa A, Nagai H, Saito M, Maruyama M, Takahashi T, Ozawa T, Tsuji S, Takahashi H (2000) An autopsy case of autosomal-recessive juvenile parkinsonism with a homozygous exon 4 deletion in the parkin gene. Mov Disord 15:884-888.
Hedrich K, Kann M, Lanthaler AJ, Dalski A, Eskelson C, Landt O, Schwinger E, Vieregge P, Lang AE, Breakefield XO, Ozelius LJ, Pramstaller PP, Klein C (2001) The importance of gene dosage studies: mutational analysis of the parkin gene in early-onset parkinsonism. Human Mol Genet 10:1649–1656.
Hedrich K, Marder K, Harris J, Kann M, Lynch T, Meija-Santana H, Pramstaller PP, Schwinger E, Bressman SB, Fahn S, Klein C (2002) Evaluation of 50 probands with early-onset Parkinson's disease for Parkin mutations. Neurology 58:1239–1246
Hicks AA, Petursson H, Jonsson T, Stefansson H, Johannsdottir HS, Sainz J, Frigge ML, Kong A, Gulcher JR, Stefansson K, Sveinbjornsdottir S (2002) A susceptibility gene for late-onset idiopathic Parkinson's disease. Ann Neurol 52:549–555.
Ibáñez P, Bonnet AM, Debarges B, Lohmann E, Tison F, Pollak P, Agid Y, Dürr A, Brice A, and the French Parkinson's Disease Genetics Study Group (2004) Alpha-synuclein gene duplication is responsible for Parkinson's disease. Lancet 364:1169–1171
Jones AC, Yamamura Y, Almasy L, Bohlega S, Elibol B, Hubble J, Kuzuhara S, Uchida M, Yanagi T, Weeks DE, Nygaard TG. Autosomal recessive juvenile parkinsonism maps to 6q25.2-q27 in four ethnic groups: detailed genetic mapping of the linked region. Am J Human Genet 1998; 63:80–87.
Kann M, Jacobs H, Mohrmann K, Schumacher K, Hedrich K, Garrels J, Wiegers K, Schwinger E, Pramstaller PP, Breakefield XO, Ozelius LJ, Vieregge P, Klein C (2002) Role of parkin mutations in 111 community-based patients with early-onset parkinsonism. Ann Neurol 51:621–625.
Khan NL, Graham E, Critchley P, Schrag AE, Wood NW, Lees AJ, Bhatia KP, Quinn N (2003) Parkin disease: a phenotypic study of a large case series. Brain 126:1279–1292.

Kitada T, Asakawa S, Hattori N, Matsumine H, Yamamura Y, Minoshima S, Yokochi M, Mizuno Y, Shimizu N (1998) Mutations in the parkin gene cause autosomal recessive juvenile parkinsonism. Nature 392:605–608.

Klein C, Pramstaller PP, Kis B, Page CC, Kann M, Leung J, Woodward H, Castellan CC, Scherer M, Vieregge P, Breakefield XO, Kramer PL, Ozelius LJ (2000) Parkin deletions in a family with adult-onset, tremor-dominant parkinsonism: expanding the phenotype. Ann Neurol 48:65–71.

Klein C, Hedrich K, Wellenbrock C, Kann M, Harris J, Marder K, Lang AE, Schwinger E, Ozelius LJ, Vieregge P, Pramstaller PP, Kramer PL (2003) Frequency of parkin mutations in late-onset Parkinson's disease. Ann Neurol 54:415–416

Kruger R, Kuhn W, Leenders KL, Sprengelmeyer R, Muller T, Woitalla D, Portman AT, Maguire RP, Veenma L, Schroder U, Schols L, Epplen JT, Riess O, Przuntek H (2001) Familial parkinsonism with synuclein pathology: clinical and PET studies of A30P mutation carriers. Neurology 56:1355–1362.

Kuroda Y, Mitsui T, Akaike M, Azuma H, Matsumoto T (2001) Homozygous deletion mutation of the parkin gene in patients with atypical parkinsonism. J Neurol Neurosurg Psychiatr 71:231–234

Lansbury, PT Jr, Brice A (2002) Genetics of Parkinson's disease and biochemical studies of implicated gene products. Cur Opin Genet Dev 12: 299–306.

Le WD, Xu P, Jankovic J, Jiang H, Appel SH, Smith RG, Vassilatis DK (2003) Mutations in NR4A2 associated with familial Parkinson disease. Nature Genet ; 33:85–89.

Leroy E, Boyer R, Auburger G, Leube B, Ulm G, Mezey E, Harta G, Brownstein MJ, Jonnalagada S, Chernova T, Dehejia A, Lavedan C, Gasser T, Steinbach PJ, Wilkinson KD, Polymeropoulos MH (1998) The ubiquitin pathway in Parkinson's disease. Nature 395:451–452.

Lincoln SJ, Maraganore DM, Lesnick TG, Bounds R, de Andrade M, Bower JH, Hardy JA, Farrer MJ (2003) Parkin variants in North American Parkinson's disease: cases and controls. Mov Disord 18:1306–1311.

Lohmann E, Periquet M, Bonifati V, Wood N, De Michele G, Bonnet AM, Fraix V, Broussolle E, Horstink MWIM, Vidailhet M, Verpillat P, Gasser T, Nicholl D, Teive H, Raskin S, Rascol O, Destée A, Ruberg M, Gasparini F, Meco G, Agid Y, Dürr A, Brice A (2003) Phenotypic variation of Parkin. Ann Neurol 54:176–185.

Lücking CB, Abbas N, Durr A, Bonifati V, Bonnet AM, de Broucker T, De Michele G, Wood NW, Agid Y, Brice A (1998) Homozygous deletions in parkin gene in European and North African families with autosomal recessive juvenile parkinsonism. The European Consortium on Genetic Susceptibility in Parkinson's Disease and the French Parkinson's Disease Genetics Study Group. Lancet 352:1355–1356.

Lücking CB, Dürr A, Bonifati V, Vaughan J, De Michele G, Gasser T, Harhangi BS, Meco G, Denefle P, Wood NW, Agid Y, Brice A, the European Consortium on Genetic Susceptibility in Parkinson's Disease and the French Parkinson's Disease Genetics Study Group. (2000) Early onset parkinsonism and parkin gene mutations. New Engl J Med 342:1560–1567.

Lücking CB, Brice A (2003) Semiquantitative PCR for the detection of exon rearrangements in the Parkin gene. Meth Mol Biol 217:13–26.

Martinez M, Brice A, Vaughan J, Zimprich A, Breteler MMB, Meco G, Filla A, Farrer M, Bétard C, hardy J, De Michele G, Bonifati V, Oostra B, Gasser T, Wood NW, Durr A (2004) Genome-wide scan linkage analysis for Parkinson's disease : the European Genetic Study of Parkinson's Disease. J Med Genet 41:900–907

Matsumine H, Saito M, Shimoda-Matsubayashi S, Tanaka H, Ishikawa A, Nakagawa-Hattori Y, Yokochi M, Kobayashi T, Igarashi S, Takano H, Sanpei K, Koike R, Mori H, Kondo T, Mizutani Y, Schaffer AA, Yamamura Y, Nakamura S, Kuzuhara S, Tsuji S, Mizuno Y. (1997) Localization of a gene for an autosomal recessive form of juvenile Parkinsonism to chromosome 6q25.2-27. Am J Human Genet 60:588–596.

Miller DW, Hague SM, Clarimon J, Baptista M, Gwinn-Hardy K, Cookson MR, Singleton AB (2004) Alpha-synuclein in blood and brain from familial Parkinson disease with SNCA locus triplication. Neurology 62:1835–1838.

Morales B, Martinez A, Gonzalo I, Vidal L, Ros R, Gomez-Tortosa E, Rabano A, Ampuero I, Sanchez M, Hoenicka J, Garcia De Yebenes J (2002) Steele-Richardson-Olszewski syndrome in a patient with a single C212Y mutation in the parkin protein. Mov Disord 17:1374–1380.

Mori H, Kondo T, Yokochi M, Matsumine H, Nakagawa-Hattori Y, Miyake T, Suda K, Mizuno Y (1998) Pathologic and biochemical studies of juvenile parkinsonism linked to chromosome 6q. Neurology 51:890–892

Muenter MD, Forno LS, Hornykiewicz O, Kish SJ, Maraganore DM, Caselli RJ, Okazaki H, Howard FM Jr, Snow BJ, Calne DB (1998) Hereditary form of parkinsonism--dementia. Ann Neurol 43:768–781.

Nichols WC, Pankratz N, Uniacke SK, Pauciulo MW, Halter C, Rudolph A, Conneally PM, Foroud T (2002) Linkage stratification and mutation analysis at the parkin locus identifies mutation positive Parkinson's disease families. J Med Genet 39:489–492

Oliveira SA, Scott WK, Martin ER, Nance MA, Watts RL, Hubble JP, Koller WC, Pahwa R, Stern MB, Hiner BC, Ondo WG, Allen FH Jr, Scott BL, Goetz CG, Small GW, Mastaglia F, Stajich JM, Zhang F, Booze MW, Winn MP, Middleton LT, Haines JL, Pericak-Vance MA, Vance JM (2003) Parkin mutations and susceptibility alleles in late-onset Parkinson's disease. Ann Neurol 53:624–629.

Oliveri RL, Zappia M, Annesi G, Bosco D, Annesi F, Spadafora P, Pasqua AA, Tomaino C, Nicoletti G, Pirritano D, Labate A, Gambardella A, Logroscino G, Manobianca G, Epifanio A, Morgante L, Savettieri G, Quattrone A (2001) The parkin gene is not involved in late-onset Parkinson's disease. Neurology 57:359–362.

Papapetropoulos S, Paschalis C, Athanassiadou A, Papadimitriou A, Ellul J, Polymeropoulos MH, Papapetropoulos T (2001) Clinical phenotype in patients with alpha-synuclein Parkinson's disease living in Greece in comparison with patients with sporadic Parkinson's disease. J Neurol Neurosurg Psychiat 70:662–665.

Periquet M, Lücking CB, Vaughan JR, Bonifati V, Dürr A, De Michele G, Horstink NW, Farrer M, Illarioshkin S, Pollak P, Borg M, Brefel-Courbon C, Denefle P, Meco G, Gasser T, Breteler MMB, Wood NW, Agid Y, Brice A, the French Parkinson s disease Genetic Study Group and the European Consortium on Genetic Susceptibility in Parkinson s disease. (2001) Origin of the mutations in the parkin gene in Europe: exon rearrangements are independent recurrent events whereas point mutations may result from founder effects. Am J Human Genet 68: 617–626.

Periquet M, Latouche M, Lohmann E, Rawal N, De Michele G, Ricard S, Teive H, Fraix V, Vidailhet M, Nicholl D, Wood NW, Raskin S, Deleuze JF, Agid Y, Dürr A, Brice A, for the French Parkinson's Disease Genetics Study Group and the European Consortium on Genetic Susceptibility in Parkinson's Disease. (2003) Parkin mutations are frequent in patients with isolated early-onset parkinsonism. Brain 126:1271–1278.

Polymeropoulos MH, Lavedan C, Leroy E, Ide SE, Dehejia A, Dutra A, Pike B, Root H, Rubenstein J, Boyer R, Stenroos ES, Chandrasekharappa S, Athanassiadou A, Papapetropoulos T, Johnson WG, Lazzarini AM, Duvoisin RC, Di Iorio G, Golbe LI, Nussbaum RL (1997) Mutation in the alpha-synuclein gene identified in families with Parkinson's disease. Science 276:2045–2047.

Pramstaller PP, Kunig G, Leenders K, Kann M, Hedrich K, Vieregge P, Goetz CG, Klein C (2002) Parkin mutations in a patient with hemiparkinsonism-hemiatrophy: a clinical-genetic and PET study. Neurology 58:808–810.

Rawal N, Periquet M, Lohmann E, Lücking CB, Teive HA, Ambrosio G, Raskin S, Lincoln S, Hattori N, Guimaraes J, Horstink MWIM, Dos Santos Bele W, Broussolle E, Destee A, Mizuno Y, Farrer M, Deleuze JF, De Michele G, Agid Y, Dürr A, Brice A, the French Parkinson s Disease Genetics Study Group and the European Consortium on Genetic Susceptibility in

Parkinson's disease. (2003) New Parkin gene mutations and atypical phenotypes in families with autosomal recessive parkinsonism. Neurology 60: 1378–1381.

Singleton AB, Farrer M, Johnson J, Singleton A, Hague S, Kachergus J, Hulihan M, Peuralinna T, Dutra A, Nussbaum R, Lincoln S, Crawley A, Hanson M, Maraganore D, Adler C, Cookson MR, Muenter M, Baptista M, Miller D, Blancato J, Hardy J, Gwinn-Hardy K (2003) alpha-Synuclein locus triplication causes Parkinson's disease. Science 302:841.

Spira PJ, Sharpe DM, Halliday G, Cavanagh J, Nicholson GA (2001) Clinical and pathological features of a Parkinsonian syndrome in a family with an Ala53Thr alpha-synuclein mutation. Ann Neurol 49:313–319.

Takahashi H, Ohama E, Suzuki S, Horikawa Y, Ishikawa A, Morita T, Tsuji S, Ikuta F (1994) Familial juvenile parkinsonism: clinical and pathologic study in a family. Neurology 44:437–441.

Tassin J, Durr A, de Broucker T, Abbas N, Bonifati V, De Michele G, Bonnet AM, Broussolle E, Pollak P, Vidailhet M, De Mari M, Marconi R, Medjbeur S, Filla A, Meco G, Agid Y, Brice A (1998) Chromosome 6-linked autosomal recessive early-onset Parkinsonism: linkage in European and Algerian families, extension of the clinical spectrum, and evidence of a small homozygous deletion in one family. The French Parkinson's Disease Genetics Study Group, and the European Consortium on Genetic Susceptibility in Parkinson's Disease. Am J Human Genet 63:88–94.

Valente EM, Abou-Sleiman PM, Caputo V, Muqit MM, Harvey K, Gispert S, Ali Z, Del Turco D, Bentivoglio AR, Healy DG, Albanese A, Nussbaum R, Gonzalez-Maldonado R, Deller T, Salvi S, Cortelli P, Gilks WP, Latchman DS, Harvey RJ, Dallapiccola B, Auburger G, Wood NW (2004) Hereditary early-onset Parkinson's disease caused by mutations in PINK1. Science 304:1158–1160.

Van De Warrenburg BPC, Lammens M, Lücking CB, Denefle P, Wesseling P, Booij J, Praamstra P, Quinn N, Brice A, Horstink MWIM (2001) Clinical and pathologic abnormalities in a family with parkinsonism and parkin gene mutations. Neurology 56: 555–557.

West A, Periquet M, Lincoln S, Lucking CB, Nicholl D, Bonifati V, Rawal N, Gasser T, Lohmann E, Deleuze JF, Maraganore D, Levey A, Wood N, Durr A, Hardy J, Brice A, Farrer M (2002) Complex relationship between Parkin mutations and Parkinson disease. Am J Med Genet 114:584–591

Yamamura Y, Sobue I, Ando K, Iida M, Yanagi T (1973) Paralysis agitans of early onset with marked diurnal fluctuation of symptoms. Neurology 23:239–244.

Zarranz JJ, Alegre J, Gomez-Esteban JC, Lezcano E, Ros R, Ampuero I, Vidal L, Hoenicka J, Rodriguez O, Atares B, Llorens V, Gomez Tortosa E, del Ser T, Munoz DG, de Yebenes JG (2004) The new mutation, E46K, of alpha-synuclein causes Parkinson and Lewy body dementia. Ann Neurol 55:164-173.

Subject Index

A
Alzheimer's disease 5–7, 12, 30, 37–42, 45–92, 108
β-amyloid 5, 6, 12, 14, 15, 39, 40, 45–55, 62, 88
amyloid protein precursor (APP) 45–51
antioxidant 7, 39, 40, 42
Apolipoprotein E polymorphism 14, 50, 62–67
Atypical clinical pictures in Alzheimer's disease 49, 52, 53

B
Biological marker 7, 40, 41, 108
Brain metabolism in frontotemporal dementia 139–151

C
Cell death 12, 13
Cerebrospinal fluid 39, 40, 42, 108
Chromosome 17-linked frontotemporal dementia with ubiquitin-positive, tau-negative inclusions 117–137
Columbian families with Alzheimer's disease 61–71
Complex disease 26–30
Cotton wool plaques 86, 87

D
Diagnosis 16, 17, 28

E
Early onset familial Alzheimer's disease 45–92
Educational level and Alzheimer's disease 65–67
Environmental risk factor 16, 61, 64–67
Ethnic influence on expression of the genotype 25–36

F
Fluctuation in cognition 5, 6
Free radical (cf. oxidative stress)
Frontotemporal dementia 3, 4, 6, 41, 42, 93–152

G
Genetic modifiers of the phenotypic presentation of Alzheimer's disease 63
Genetic of Alzheimer's disease 12, 14–16, 30, 45–92, 93, 94, 99
Genetic of amyotrophic lateral sclerosis 13
Genetic of frontotemporal dementia 28, 45, 49, 54, 55, 93–152
Genetic of Parkinson's disease 12–16, 29, 153–164
Genetic of polyglutamine expansion disorders 13
Genetic of prion disorders 14, 15
Genetic of spinocerebellar ataxia 16, 29–31
Genetic transmission of neurodegeneratives disorders 3, 11–17, 25–36

H
4-hydroxy-2-nonenal 39, 41

I
Inherited and sporadic forms of neurodegenerative disorders 3, 11–17
Isoprostane 39–42

L
Language disorder 5
Large familial Alzheimer's disease kindred 61–71

M
Memory disorder 5, 6
Mendelian form of disease 12–15, 26
Mild cognitive impairment 7, 40
Molecular classification of neurodegenerative disease 11–23
Molecular genetic dissection of disease etiology 11–15

Monogenic neurodegenerative diseases 29–32, 74, 154
Mutation and prediction of pathology 45–59, 93–101

O

Oxidative stress in neurodegenerative diseases 37–44

P

Parkin gene mutation 154, 156–160
Parkinson's disease 4–7, 17, 30, 33, 153–164
PET scan (in frontotemporal dementia) 139–150
Phenotype/genotype correlations in Parkinson's disease 153–164
Presenilin 45–89, 93–101
Proteinopathie (neurodegenerative disorder as) 1–10
Proteomic analysis 7

R

Race (definition) 25, 26
Racial influence on expression of the genotype in neurodegenerative diseases 25–36

S

Sleep disorder 5
Spastic paraparesis 52–54, 73–92
Spinocerebellar ataxia 29–31
α-synucléinopathies 4–7, 12, 30, 154–156

T

Tauopathies 3, 4, 12, 14, 16, 41, 104, 106–110, 118
Treatment 6, 7, 16, 17, 32, 39, 40

U

Ubiquitin-positive inclusions 117–137

V

Variable phenotype of Alzheimer's disease with spastic paraparesis 73–92

MIX
Papier aus verantwortungsvollen Quellen
Paper from responsible sources
FSC® C105338

If you have any concerns about our products,
you can contact us on
ProductSafety@springernature.com

In case Publisher is established outside the EU,
the EU authorized representative is:
Springer Nature Customer Service Center GmbH
Europaplatz 3, 69115 Heidelberg, Germany

Printed by Libri Plureos GmbH
in Hamburg, Germany